UNIVERSITY OF NOTTINGHAM
D0177572

2011

ONE WEEK LOAN
ONE WEEK ONLY
26 FEB 1993
24 MAY 1993
1993
ONE WEEK ONLY
2 JUN 1993
12 MAR 1993

Electromagnetism for Electronic Engineers

TUTORIAL GUIDES IN ELECTRONIC ENGINEERING

This series is aimed at first- and second-year undergraduate courses. Each text is complete in itself, although linked with others in the series. Where possible, the trend towards a 'systems' approach is acknowledged, but classical fundamental areas of study have not been excluded. Worked examples feature prominently and indicate, where appropriate, a number of approaches to the same problem.

A format providing marginal notes has been adopted to allow the authors to include ideas and material to support the main text. These notes include references to standard mainstream texts and commentary on the applicability of solution methods, aimed particularly at covering points normally found difficult. Graded problems are provided at the end of each chapter, with answers at the end of the book.

1. Transistor Circuit Techniques: Discrete and integrated (2nd edn) – G.J. Ritchie
2. Feedback Circuits and Op Amps (2nd edn) – D.H. Horrocks
3. Pascal for Electronic Engineers (2nd edn) – J. Attikiouzel
4. Computers and Microprocessors: Components and systems (3rd edn) – A.C. Downton
5. Telecommunication Principles (2nd edn) – J.J. O'Reilly
6. Digital Logic Techniques: Principles and practice (2nd edn) – T.J. Stonham
7. Instrumentation: Transducers and Interfacing (2nd edn) – B.R. Bannister and D.G. Whitehead
8. Signals and Systems: Models and behaviour (2nd edn) – M.L. Meade and C.R. Dillon
9. Basic Electromagnetism and its Applications – A.J. Compton
10. Electromagnetism for Electronic Engineers (2nd edn) – R.G. Carter
11. Power Electronics – D.A. Bradley
12. Semiconductor Devices: How they work – J.J. Sparkes
13. Electronic Components and Technology: Engineering applications – S.J. Sangwine
14. Optoelectronics – J. Watson
15. Control Engineering – C. Bissell
16. Basic Mathematics for Electronic Engineers: Models and applications – Szymanski
17. Software Engineering – D. Ince
18. Integrated Circuit Design and Technology – M.J. Morant

Electromagnetism for Electronic Engineers

Second edition

R.G. Carter
Department of Engineering
University of Lancaster

CHAPMAN & HALL
University and Professional Division
London · Glasgow · New York · Tokyo · Melbourne · Madras

Published by Chapman & Hall, 2–6 Boundary Row, London SE1 8HN

Chapman & Hall, 2–6 Boundary Row, London SE1 8HN, UK

Blackie Academic & Professional, Wester Cleddens Road, Bishopbriggs, Glasgow G64 2NZ, UK

Van Nostrand Reinhold Inc., 115 5th Avenue, New York NY10003, USA

Chapman & Hall Japan, Thomson Publishing Japan, Hirakawacho Nemoto Building, 6F, 1-7-1 Hirakawa-cho, Chiyoda-ku, Tokyo 102, Japan

Chapman & Hall Australia, Thomas Nelson Australia, 102 Dodds Street, South Melbourne, Victoria 3205, Australia

Chapman & Hall India, R. Seshadri, 32 Second Main Road, CIT East, Madras 600 035, India

First edition 1986
Reprinted 1987, 1989
Second edition 1992

Typeset in Times 10 on 12 pt by Colset Private Limited, Singapore
Printed and bound in Hong Kong

ISBN 0 412 42740 0 0 442 31558 9 (USA)

A catalogue record for this book is available from the British Library

Library of Congress Cataloging-in-Publication data available

To my father, Geoffrey William Carter (1909–1989), a distinguished teacher of electromagnetic theory

Contents

Preface x

Preface to the first edition xi

1 Electrostatics in free space 1

The inverse square law of force between two electric charges 1
The electric field 2
Gauss' theorem 4
The differential form of Gauss' theorem 6
Electrostatic potential 7
Calculation of potential in simple cases 9
Calculation of the electric field from the potential 10
Conducting materials in electrostatic fields 11
Methods of solving electrostatic field problems 13
The method of images 14
Laplace's and Poisson's equations 16
The finite difference method 17
The motion of charged particles in electric fields 19

2 Dielectric materials and capacitance 23

Insulating materials in electric fields 23
Solution of problems involving dielectric materials 25
Boundary conditions 26
Capacitance 28
Electrostatic screening 29
Calculation of capacitance 31
Energy storage in the electric field 35
Calculation of capacitance by energy methods 37
Finite element methods 38
Boundary element methods 39

3 Steady electric currents 43

Conduction of electricity 43
Ohmic heating 44
The distribution of current density in conductors 45
Electric fields in the presence of currents 47
Electromotive force 48
Calculation of resistance 49
Calculation of resistance by energy methods 51

4 **The magnetic effects of electric currents** 55

The law of force between two moving charges 55
The magnetic flux density 56
The magnetic circuit law 59
Magnetic scalar potential 60
Forces on charges moving in magnetic fields 61
Motion of charges in combined electric and magnetic fields 63
Forces on current-carrying conductors 64

5 **The magnetic effects of iron** 67

Ferromagnetic materials 68
Boundary conditions 70
Flux conduction and magnetic screening 71
Magnetic circuits 74
Fringing and leakage 75
Hysteresis 78
Solution of problems in which μ cannot be regarded as constant 80
Permanent magnets 83
Using permanent magnets efficiently 85

6 **Electromagnetic induction** 90

The current induced in a conductor moving through a steady magnetic field 90
The current induced in a loop of wire moving through a non-uniform magnetic field 92
Faraday's law of electromagnetic induction 93
Inductance 94
Electromagnetic interference 96
Calculation of inductance 98
Energy storage in the magnetic field 101
Calculation of inductance by energy methods 103
The *LCRZ* analogy 105
Energy storage in iron 107
Hysteresis loss 108
Eddy currents 109
Real electronic components 110

7 **Transmission lines** 113

The circuit theory of transmission lines 113
Representation of waves using complex numbers 116
Characteristic impedance 117
Reflection of waves at the end of a line 117
Pulses on transmission lines 119
Reflection of pulses at the end of a line 120

Transformation of impedance along a transmission line 124
The quarter-wave transformer 126
Field description of transmission lines 127
The electric and magnetic fields in a coaxial line 128
Power flow in a coaxial line 129

8 Maxwell's equations and electromagnetic waves 133

Maxwell's form of the magnetic circuit law 133
The differential form of the magnetic circuit law 135
The differential form of Faraday's law 136
Maxwell's equations 137
Plane electromagnetic waves in free space 138
Power flow in an electromagnetic wave 139

9 Screening circuits against radio-frequency interference 142

The small electric dipole 143
The small magnetic dipole 146
Electromagnetic waves in conducting materials 148
Reflection of electromagnetic waves by a conducting surface 151
The effect of holes in screening enclosures 153
Electromagnetic resonances 157
The effect of resonance on screening effectiveness 159

Bibliography 164

Appendix Physical constants 166

Properties of dielectric materials 166

Properties of conductors 166

Properties of ferromagnetic materials 167

Summary of vector formulae in Cartesian coordinates 167

Summary of the principal formulae of electromagnetism 167

Answers to problems 170

Index 172

Preface

A further six years, or so, of teaching and thinking about electromagnetic theory have confirmed my belief that this subject is a vital part of the education of electronic engineers. Without it they are limited to understanding electronic circuits in terms of the idealizations of circuit theory. In this second edition I have tried to strengthen still further the sections which deal with the non-ideal behaviour of electronic components and circuits.

I have also taken the opportunity to add a chapter discussing the applications of electromagnetic theory to electromagnetic compatibility (emc). The imminent adoption of stringent EC standards for emc makes some familiarity with the subject essential for all graduate electronic engineers. It has been necessary in this treatment to quote without proof many of the equations required. I hope that this will be sufficient for those whose studies of electromagnetic theory will not go beyond this book. Those who do proceed to more advanced studies will find the proofs in my book *Electromagnetic Waves: Microwave Components and Devices* (Chapman & Hall, 1990). I am grateful to Dr Marvin of York University who read and commented on this chapter in draft.

Finally, I have taken this opportunity to correct a number of mistakes in the first edition. I am particularly indebted to Professor Freeman of Imperial College, London, and Dr Sykulski of Southampton University for pointing out mistakes in my discussion of energy methods.

Preface to the first edition

Electromagnetism is fundamental to the whole of electrical and electronic engineering. It provides the basis for the understanding of the uses of electricity and for the design of the whole spectrum of devices from the largest turbo-alternators to the smallest microcircuits. Every electrical or electronic engineer ought to have a sound knowledge of the subject so that he knows why, as well as how, the components which he uses differ from the idealizations of circuit theory. Many students of electronic engineering could benefit from a greater understanding of the range of electronic components. Besides the familiar circuit components there are special components which are used at high frequencies or high powers, transducers for turning information into electrical signals, and devices such as loudspeakers, recording heads and displays for turning electronic signals into other kinds of energy. Electromagnetic theory is an essential tool for those engineers who design and develop such electronic components and devices.

This book is one of a series of tutorial texts which aim to make their subjects particularly accessible to students, and its title is intended to be an accurate description of the aims of this particular volume in the series.

The book is, first and foremost, about electromagnetism, and any book which covers this subject must deal with its various laws. But electromagnetism to some extent involves a circular argument: you can choose different ways of entering its description and still, in the end, cover the same ground. I have chosen a conventional sequence of presentation, beginning with electrostatics, then moving to current electricity, the magnetic effects of currents, electromagnetic induction and electromagnetic waves. This seems to me to be the most logical approach. Authors differ also in the significance they ascribe to the four field vectors $\boldsymbol{E}$, $\boldsymbol{D}$, $\boldsymbol{B}$ and $\boldsymbol{H}$. I find it simplest to regard $\boldsymbol{E}$ and $\boldsymbol{B}$ as 'physical' quantities because they are directly related to forces on charged particles, and $\boldsymbol{D}$ and $\boldsymbol{H}$ as useful inventions which make it easier to solve problems involving material media. For this reason the introduction of $\boldsymbol{D}$ and $\boldsymbol{H}$ is deferred until the points at which they are needed for this purpose.

Secondly, this is a book for those whose main interest is in electronics. The restricted space available meant that decisions had to be taken about what to include or omit. I have chosen the topics for the worked examples and the problems exclusively from electronic engineering. Where topics, such as the force on an iron surface in a magnetic field, have been omitted, it is because they are of marginal importance in electronics.

Thirdly, I have written a book for engineers. On the whole engineers can take the laws of physics as read. Their task is to apply them to the practical problems they meet in their work. For this reason I have chosen to introduce the laws with demonstrations of plausibility rather than formal proofs. It seems to me that engineers understand things best from practical examples rather than abstract mathematics. I have found from experience that few textbooks on electromagnetism are much help when it comes to applying the subject, so here I have tried to make good that deficiency both by emphasizing the strategies of problem-solving and the range of techniques available and by giving, through the marginal notes, a way into the professional literature. One point deserves special mention. I have included an introduction to the use of energy methods for solving electromagnetic problems. These methods are usually only encountered in graduate-level texts or the professional literature. But they can be explained very simply and they provide the student with a means of estimating circuit parameters quite simply to an accuracy which is good enough for most purposes. As far as I know, no other undergraduate text includes them.

Most university engineering students already have some familiarity with the fundamentals of electricity and magnetism from their school physics courses. This book is designed to build on that foundation by providing a systematic treatment of a subject which may previously have been encountered as a set of experimental phenomena with no clear links between them. Those who have not studied the subject before, or who feel a need to revise the basic ideas, should consult the companion volume in this series by Tony Compton (*Basic Electromagnetism and its Applications*, Van Nostrand Reinhold (1986)). That book is also useful in that it approaches the subject in a somewhat different way, giving an alternative perspective upon it. In particular, it gives rather more emphasis to the properties of materials and to their application in a wide range of devices than this book.

The mathematical techniques used in this book are all covered either at A level or during the first year at university. They include calculus, coordinate geometry and vector algebra, including the use of dot and cross products. Vector notation makes it possible to state the laws of electromagnetism in concise general forms. This advantage seems to me to outweigh the possible disadvantage of its relative unfamiliarity. I have introduced the notation of vector calculus in order to provide students with a basis for understanding more advanced texts which deal with electromagnetic waves. No attempt is made here to apply the methods of vector calculus because the emphasis is on practical problem-solving and acquiring insight and not on the application of advanced mathematics.

I am indebted for my understanding of this subject to many people, teachers, authors and colleagues, but I feel a particular debt to my father who taught me the value of thinking about problems 'from first principles'. His own book, *The Electromagnetic Field in its Engineering Aspects* (2nd edn, Longman, 1967) is a much more profound treatment than I have been able to attempt, and is well worth consulting. I should like to record my gratitude to my editors, Professors Bloodworth and Dorey, of the white and the red roses, to Tony Compton and my colleague David Bradley, all of whom read the draft of the book and offered many helpful suggestions. Finally, I now realize why authors acknowledge the support and forbearance of their wives and families through the months of burning the midnight oil, and I am most happy to acknowledge my debt there also.

Electrostatics in free space 1

Objectives

- ☐ To show how the idea of the electric field is based on the inverse square law of force between two electric charges.
- ☐ To explain the principle of superposition and the circumstances in which it can be applied.
- ☐ To explain the concept of the flux of an electric field.
- ☐ To introduce Gauss' theorem and to show how it can be applied to those cases where the symmetry of the problem makes it possible.
- ☐ To derive the differential form of Gauss' theorem.
- ☐ To introduce the concept of electrostatic potential difference and to show how to calculate it from a given electric field distribution.
- ☐ To show how the velocities of charged particles can be calculated using the principle of conservation of energy.
- ☐ To explain the idea of the gradient of the potential and to show how it can be used to calculate the electric field from a given potential distribution.
- ☐ To show how flux plots can be sketched for simple arrangements of charges and electrodes.
- ☐ To show how simple problems involving electrodes with applied potentials can be solved using Gauss' theorem, the principle of superposition and the method of images.
- ☐ To introduce the Laplace and Poisson equations.
- ☐ To show how the finite difference method can be used to find the solution to Laplace's equation for simple two-dimensional problems.
- ☐ To discuss the types of problem in which Poisson's equation must be solved.
- ☐ To show how the motion of charged particles in electric fields can be calculated in simple cases.

The inverse square law of force between two electric charges

The idea that electric charges exert forces on each other needs no introduction to anyone who has ever drawn a comb through his or her hair and used it to pick up small pieces of paper. The existence of electric charges and of the forces between them underlies every kind of electrical or electronic device. For the present we shall concentrate on the forces between charges which are at rest and on the force exerted on a moving charge by other charges which are at rest. The question of the forces between moving charges is a little more difficult. We shall return to it in Chapter 4, where it is shown that magnetic effects can be regarded as due to the motion of electric charges. The science of phenomena involving stationary electric charges, known as electrostatics, finds many applications in electronics, including the calculation of capacitance and the theory of every type of active electronic device. Electrostatic phenomena are put to work in electrostatic copiers and paint sprays. They can also be a considerable nuisance, leading to explosions in oil tankers and the need for special precautions when handling metal-oxide semiconductor integrated circuits.

The experimental proof of the inverse square law involves showing that there is no electric field within a closed conducting shell. The method and the underlying theory are described by Bleaney and Bleaney.*

The starting point for the discussion of electrostatics is the experimentally determined law of force between two concentrated charges. This law, first established by Coulomb (1785), is that the force is proportional to the product of the magnitudes of the charges and inversely proportional to the square of the distance between them. Subsequent experiments have shown that the power involved cannot deviate from 2 by more than one part in 10^9. In the shorthand of mathematics the law may be written

The factor of 4π which appears on the bottom line of Equation (1.1) is the result of a decision to 'rationalize' the system of units. The effect is that π appears in those formulae which have cylindrical or spherical symmetry.

$$F = \frac{Q_1 Q_2}{4\pi\epsilon_0 r^2}\hat{r} \tag{1.1}$$

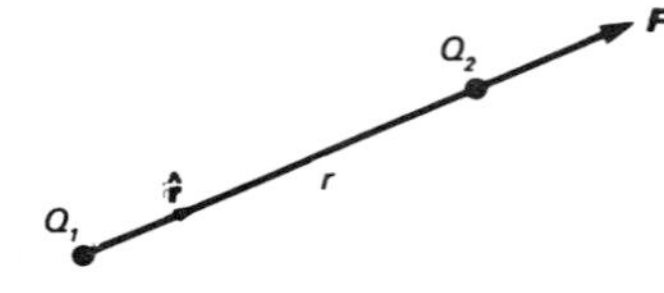

where Q_1 and Q_2 are the magnitudes of the two charges and r is the distance between them, as shown in the figure in the margin. Now force is a vector quantity, so Equation (1.1) includes the unit vector $\hat{r}$ which is directed from Q_1 towards Q_2 and the equation gives the force exerted on Q_2 by Q_1. The force exerted on Q_1 by Q_2 is equal and opposite, as required by Newton's third law of motion.

Examination of Equation (1.1) shows that it includes the effect of the polarity of the charges correctly, so that like charges repel each other while unlike charges attract. The symbol ϵ_0 denotes the **primary electric constant**; its value depends upon the system of units being used. In this book SI units are used throughout, as is now the almost universal practice of engineers. In this system of units ϵ_0 is measured in farads per metre, and its experimental value is $8.854 \times 10^{-12}\,\mathrm{F\,m^{-1}}$; the unit of charge is the **coulomb** (C).

A table giving the values of the constants introduced in this book is given in Appendix 1.

ϵ_0 is also sometimes known as the permittivity of free space.

Electric charge on a macroscopic scale is the result of the accumulation of large numbers of atomic charges each having magnitude 1.602×10^{-19} C. These charges may be positive or negative, protons being positively charged and electrons negatively. In nearly all problems in electronics the electrons are movable charges while the protons remain fixed in the crystal lattices of solid conductors or insulators. The exceptions to this occur in conduction in liquids and gases, where positive ions may contribute to the electric current.

The mobile positive charges known as 'holes' which occur in semiconducting materials are the result of quantum-mechanical effects involving moving electrons. The details can be found in Bar-Lev.*

The electric field

Although Equation (1.1) is fundamental to the theory of electrostatics it is seldom, if ever, used directly. The reason for this is that we are usually interested in effects involving large numbers of charges, so that the use of Equation (1.1) would require some sort of summation over the (vector) forces on a charge produced by every other charge. This is not normally easy to do and, as we shall see later, the distribution of charges is not always known, though it can be calculated if necessary. Equation (1.1) can be divided into two parts by the introduction of a new vector $\boldsymbol{E}$, so that

There are different ways of introducing the vectors $\boldsymbol{E}$ and $\boldsymbol{D}$ which describe the electric field. Compton* uses a different approach. You must make up your own mind about which you find most helpful.

$$E = \frac{Q_1}{4\pi\epsilon_0 r^2}\hat{r} \tag{1.2}$$

Note: The asterisks (*) in the marginal notes identify books listed in the Bibliography on page 164.

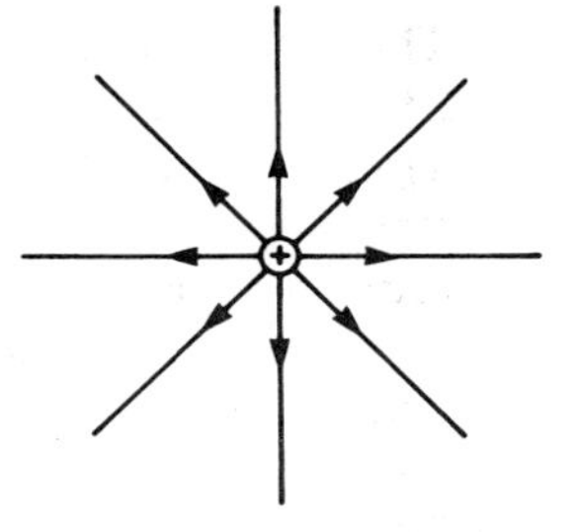

Fig. 1.1 The electric field of a point charge can be represented diagrammatically by lines of force. The figure should really be three-dimensional, with the lines distributed evenly in all directions.

$$\text{and } \boldsymbol{F} = Q_2 \boldsymbol{E} \tag{1.3}$$

The vector $\boldsymbol{E}$ is known as the **electric field**, and is measured in volts per metre in SI units. The step of introducing $\boldsymbol{E}$ is important because it separates the source of the electric force (Q_1) from its effect on the charge Q_2. The question of whether the electric field has a real existence or not is one which we can leave to the philosophers of science; its importance to engineers is that it is an effective tool for solving problems. The electric field is often represented by diagrams like Fig. 1.1 in which the lines, referred to as 'lines of force', show the direction of $\boldsymbol{E}$. The arrowheads show the direction of the force which would act on a positive charge placed in the field. The spacing of the lines of force is inversely proportional to the magnitude of the field. This kind of diagram is a useful aid to thought about electric fields, so it is well worth while becoming proficient in sketching the field patterns associated with different arrangements of charges. We shall return to this point later, when discussing electric fields in the presence of conducting materials.

In order to move from the idea of the force acting between two point charges to that acting on a charge due to a whole assembly of other charges it is necessary to invoke the **principle of superposition**. This principle applies to any linear system, that is, one in which the response of the system is directly proportional to the stimulus producing it. The principle states that the response of the system to a set of stimuli applied simultaneously is equal to the sum of the responses produced when the stimuli are applied separately. Equation (1.2) shows that the electric field in the absence of material media ('in free space') is proportional to the charge producing it, so the field produced by an assembly of charges is the vector sum of the fields due to the individual charges. The principle of superposition is very valuable because it allows us to tackle complicated problems by treating them as the sums of simpler problems. It is important to remember that the principle can be applied only to linear systems. The response of some materials to electric fields is non-linear and the use of the principle is not valid in problems involving them.

Other examples of linear systems are electric networks made up of linear components, and elastic structures.

Before discussing ways of calculating the electric field it is worth noting why we might wish to do it. The information might be needed to calculate:

- the forces on charges;
- the conditions under which voltage breakdown would occur;
- capacitance;
- the electrostatic forces on material media.

Further details of these applications of electrostatics can be found in Compton* and in Lorrain P. and Corson D.R., *Electromagnetism, Principles and Applications*, Freeman (1978).

The last of these is of limited importance because the forces are tiny. They are put to use in electrostatic loudspeakers, copiers, ink-jet printers and paint sprays.

Gauss' theorem

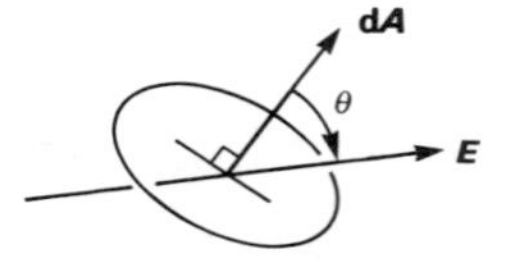

Figure 1.1 shows electric field lines radiating from a charge in much the same way that flow lines in an incompressible fluid radiate from a source (such as the end of a thin pipe) immersed in a large volume of fluid (shown in Fig. 1.2). Now in the fluid the volume flow rate across a control surface such as S, which encloses the end of the pipe, must be independent of the surface chosen and equal to the flow rate down the pipe, that is, to the strength of the source.

To apply this idea to the electric field it is necessary to define the equivalent of the flow rate which is known as the **electric flux**. The figure in the margin shows a small element of surface of area $\mathrm{d}A$ and the local direction of the electric field $\boldsymbol{E}$. The flux of $\boldsymbol{E}$ through $\mathrm{d}A$ is defined as the product of the area with the normal component of $\boldsymbol{E}$. This can be written very neatly using vector notation by defining a vector $\mathbf{d}\boldsymbol{A}$ normal to the surface element. The flux of $\boldsymbol{E}$ through $\mathbf{d}\boldsymbol{A}$ is then just $\boldsymbol{E}\cdot\mathbf{d}\boldsymbol{A} = E\mathrm{d}A\cos\theta$.

The symbol $\boldsymbol{A}$ is also used for magnetic vector potential. That concept is not used in this book so the risk of confusion is small.

Now consider the total flux coming from a point charge. The simplest choice of control surface (usually called a **Gaussian surface** in this context) is a sphere concentric with the charge. Equation (1.2) shows that $\boldsymbol{E}$ is always normal to the surface of the sphere and its magnitude is constant there. This makes the calculation of the flux of $\boldsymbol{E}$ out of the sphere easy — it is just the product of the magnitude of $\boldsymbol{E}$ with the surface area of the sphere:

$$\text{flux of } \boldsymbol{E} = \frac{Q}{4\pi\epsilon_0 r^2}4\pi r^2 = \frac{Q}{\epsilon_0} \tag{1.4}$$

Thus the flux of $\boldsymbol{E}$ out of the sphere is independent of the radius of the sphere and depends only on the charge enclosed within it. It can be shown that this result is true for any shape of surface and, by using the principle of superposition, for any grouping of charges enclosed. The result may be stated in words:

The flux of $\boldsymbol{E}$ in free space out of any closed surface is equal to the charge enclosed by the surface divided by ϵ_0.

The proof of Gauss' theorem for any shape of surface can be found in Bleaney and Bleaney.*

Remember that this form of Gauss' theorem is valid only for charges in free space. A modified form which is used when material media are present is discussed in the next chapter.

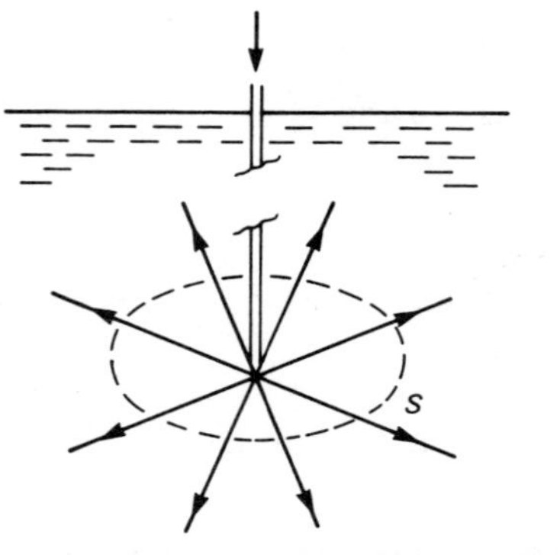

Fig. 1.2 The flow of an incompressible fluid from the end of a thin pipe is analogous to the electric field of a point charge. It is necessary for the end of the pipe to be well away from the surface of the fluid and the walls of the containing vessel.

This is known as Gauss' theorem. It can also be written, using the notation of mathematics, as

$$\oint\!\!\oint_S \boldsymbol{E}\cdot\mathbf{d}\boldsymbol{A} = \frac{1}{\epsilon_0}\iiint_v \rho \mathrm{d}v \qquad (1.5)$$

The symbol $\oint\!\!\oint$ is used to show that the integral is to be taken over a closed surface.

where S is a closed surface enclosing the volume v and ρ is the charge density. Equation (1.5) looks fearsome but, in fact, it is possible to apply it directly only in the very limited range of cases whose symmetry allows the integrals to be evaluated.

Worked Example 1.1

Figure 1.3 shows a cylinder having charge q per unit length distributed uniformly over its surface. Find an expression for the electric field at any point outside the cylinder.

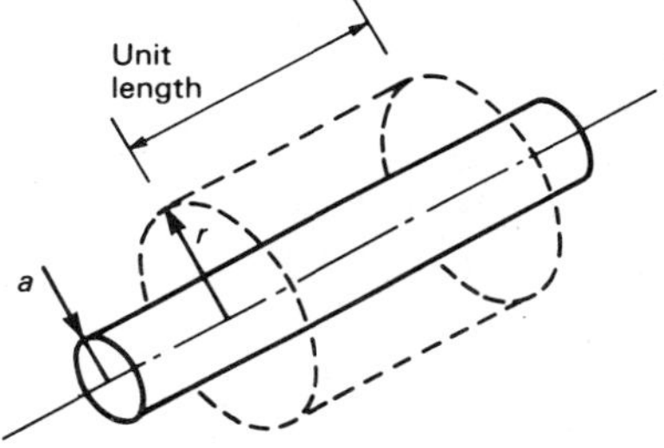

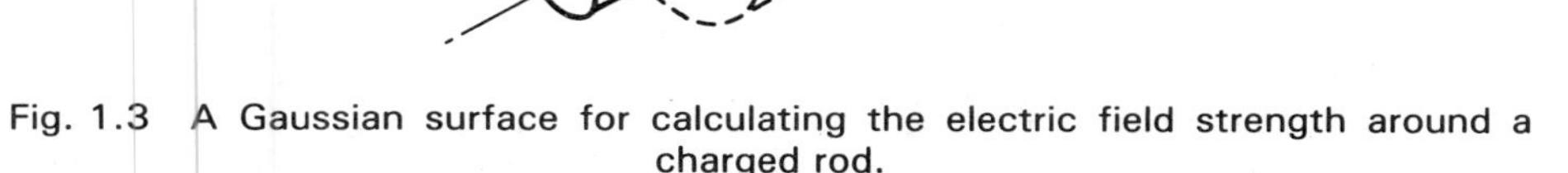

Fig. 1.3 A Gaussian surface for calculating the electric field strength around a charged rod.

Solution From the symmetry of the problem we can assume that $\boldsymbol{E}$ is everywhere directed radially outwards and that the magnitude of $\boldsymbol{E}$ depends only on the distance from the axis. This is not valid near the ends of the cylinder but the problem can be solved in this way only if this assumption is made. The next step is to define the Gaussian surface to be used. This is chosen to be a cylinder of radius r and unit length concentric with the charged cylinder with ends which are flat and perpendicular to the axis. On the curved part of the Gaussian surface $\boldsymbol{E}$ has constant magnitude and is everywhere perpendicular to the surface. The flux of $\boldsymbol{E}$ out of this part of the surface is therefore equal to the product of E and the area of the curved surface. On the ends of the cylinder $\boldsymbol{E}$ is not constant but, since it is always parallel to the surface, the flux of $\boldsymbol{E}$ out of the ends of the cylinder is zero. Finally, since the Gaussian surface is of unit length it encloses charge q. Therefore, from Gauss' theorem,

This method gives the magnitude of $\boldsymbol{E}$, written E. Its direction is deduced from the symmetry of the problem.

$$2\pi rE = q/\epsilon_0$$

or

$$E = q/2\pi\epsilon_0 r \qquad (1.6)$$

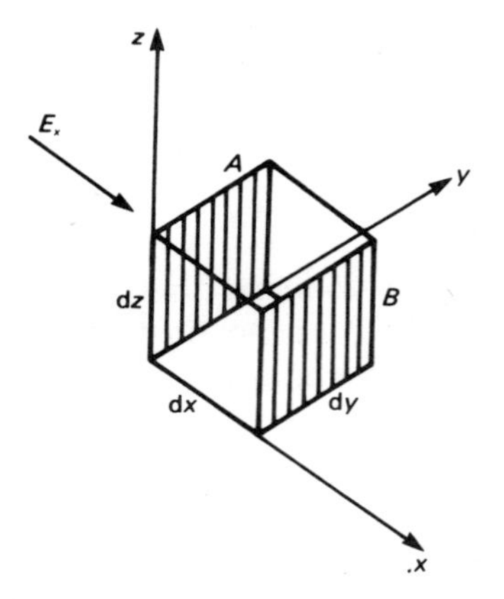

Fig. 1.4 The elementary Gaussian surface used to derive the differential form of Gauss' theorem.

The differential form of Gauss' theorem

Only a limited range of problems can be solved by the direct use of Equation (1.5). Another form, which is obtained by applying it to a small volume element, enables us to solve a much wider range of problems. Figure 1.4 shows such a volume element in Cartesian coordinates. To calculate the net flux out of the element, consider first the two shaded faces A and B which are perpendicular to the x-axis. The only component of $\boldsymbol{E}$ which contributes to the flux through these faces is E_x. If the component of $\boldsymbol{E}$ is E_x on A then, in general, the x component of $\boldsymbol{E}$ on B can be written $E_x + (\partial E_x/\partial x)\,\mathrm{d}x$. Provided that the dimensions of the element are small enough we can assume that these components are constant on the surfaces A and B. The flux of $\boldsymbol{E}$ out of the volume through the faces A and B is then

The mathematical background to this section can be found in Kreyszig E., *Advanced Engineering Mathematics* (5th edn), Wiley (1983). The same book gives the appropriate forms for expressions involving cylindrical and spherical polar coordinates. A summary of the vector formulae used in this book is given in the Appendix.

$$-E_x\,\mathrm{d}y\,\mathrm{d}z + \left(E_x + \frac{\partial E_x}{\partial x}\,\mathrm{d}x\right)\mathrm{d}y\,\mathrm{d}z = \frac{\partial E_x}{\partial x}\,\mathrm{d}x\,\mathrm{d}y\,\mathrm{d}z$$

The same argument can be used for the other two directions in space, with the result that the net flux of $\boldsymbol{E}$ out of the element is

$$\left(\frac{\partial E_x}{\partial x} + \frac{\partial E_y}{\partial y} + \frac{\partial E_z}{\partial z}\right)\mathrm{d}x\,\mathrm{d}y\,\mathrm{d}z \tag{1.7}$$

Now if ρ is the local charge density, which may be assumed to be constant if the volume element is small enough, the charge enclosed in the volume is

$$\rho\,\mathrm{d}x\,\mathrm{d}y\,\mathrm{d}z \tag{1.8}$$

Applying Gauss' theorem to the element and making use of Equations (1.7) and (1.8) gives the differential form of Gauss' theorem:

This equation applies only in free space. When dielectric materials are present Equation (2.6) must be used.

$$\frac{\partial E_x}{\partial x} + \frac{\partial E_y}{\partial y} + \frac{\partial E_z}{\partial z} = \rho/\epsilon_0 \tag{1.9}$$

The expression on the left-hand side of Equation (1.9) is known as the **divergence** of $\boldsymbol{E}$. It is sometimes written as div $\boldsymbol{E}$. The same expression can also be written as the dot product between the differential operator

The term divergence comes from the use of similar mathematical methods in the theory of fluid flow. An incompressible fluid flow has zero divergence in a region of space which is free from sources or sinks of the flow.

$$\nabla = \left(\hat{\boldsymbol{x}}\frac{\partial}{\partial x} + \hat{\boldsymbol{y}}\frac{\partial}{\partial y} + \hat{\boldsymbol{z}}\frac{\partial}{\partial z}\right) \tag{1.10}$$

and the vector $\boldsymbol{E}$. In Equation (1.10) $\hat{\boldsymbol{x}}$, $\hat{\boldsymbol{y}}$, $\hat{\boldsymbol{z}}$ are unit vectors along the x-, y- and z-axes. Using the symbol ∇, which is known as 'del', Equation (1.9) can be written

$$\nabla \cdot \boldsymbol{E} = \rho/\epsilon_0 \tag{1.11}$$

This abbreviation is not as pointless as it seems because Equation (1.11) is valid for all systems of coordinates in which the coordinate surfaces intersect at right angles. An appropriate form for ∇ can be found for each such coordinate system.

Electrostatic potential

The electric field is inconvenient to work with because it is a vector; it would be much simpler to be able to work with scalar variables. The **electrostatic potential difference** (V) between two points in an electric field is defined as the work done when unit positive charge is moved from one point to the other. Consider the figure in the margin. The force on the charge is $\boldsymbol{E}$, from Equation (1.3), so the external force needed to hold it in equilibrium is $-\boldsymbol{E}$. The work done on the charge by the external force when it is moved through a small distance $\mathbf{d}\boldsymbol{l}$ is the product of the external force and the distance moved in the direction of that force. Thus the change in electrostatic potential is

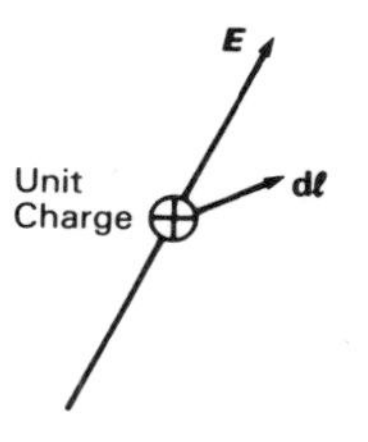

$$\mathrm{d}V = -\boldsymbol{E}\cdot\mathbf{d}\boldsymbol{l} \tag{1.12}$$

The potential difference between two points A and B can be calculated by integrating Equation (1.12) along any path between them. Mathematically this is written

$$V_B - V_A = -\int_A^B \boldsymbol{E}\cdot\mathbf{d}\boldsymbol{l} \tag{1.13}$$

This kind of integral is called a **line integral**. This is a slightly tricky concept, but its application is limited in practice to cases where the symmetry of the problem makes its evaluation possible.

The integral forms of the laws of electromagnetism are the simplest to understand, but the differential forms derived from them are much more useful for solving problems.

The electrostatic potential is analogous to gravitational potential, which is defined as the work done in moving a unit mass against gravity from one point to another. The change in the gravitational potential depends only upon the relative heights of the starting and finishing points and not on the path which is taken between them. We can show that the same is true for the electrostatic potential. Figure 1.5 shows a possible path between two points A and B in the presence of the electric field due to a point charge Q at O. The contribution to the integral of Equation (1.13) from a small movement $\mathbf{d}\boldsymbol{l}$ of a unit charge at P is

$$\mathrm{d}V = -\frac{Q}{4\pi\epsilon_0 r^2}\hat{\boldsymbol{r}}\cdot\mathbf{d}\boldsymbol{l} \tag{1.14}$$

But the dot product $\hat{\boldsymbol{r}}\cdot\mathbf{d}\boldsymbol{l}$ is simply a way of writing 'the component of $\mathbf{d}\boldsymbol{l}$ in the

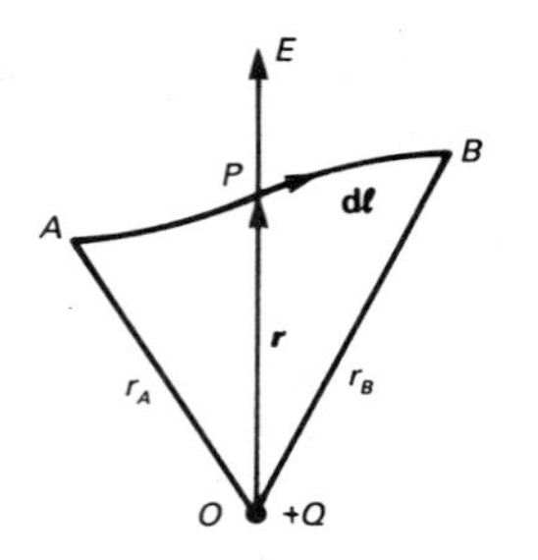

Fig. 1.5 When a charge is moved from A to B in the field of another charge at O the change in electrostatic potential is found to be independent of the choice of the path APB. It depends only on the positions of the ends of the path.

radial direction' using mathematical notation, and this quantity is just dr, the change in the distance from O. Thus Equation (1.14) can be integrated to give

$$V_B - V_A = -\int_A^B \frac{Q}{4\pi\epsilon_0 r^2}\,\mathrm{d}r = \frac{Q}{4\pi\epsilon_0}\left(\frac{1}{r_B} - \frac{1}{r_A}\right) \tag{1.15}$$

The potential difference between A and B therefore depends only on their positions and not on the path taken between them. By using the principle of superposition we can extend this proof to the field of any combination of charges.

Potential differences are measured in volts. They are familiar to electronic engineers from their role in the operation of electronic circuits. It is important to remember that potentials are always relative. Any convenient point can be chosen as the zero of potential to which all other voltages are referred. Electronic engineers are inclined to speak loosely of the voltage at a point in a circuit when strictly they mean the voltage relative to the common rail. It is as well to keep this point in mind.

It follows from the preceding discussion that the line integral of the electric field around a closed path is zero. This is really a formal way of saying that the principle of conservation of energy applies to the motion of charged particles in electric fields. In mathematical symbols the line integral around a closed path is indicated by adding a circle to the integral sign so that

$$\oint \boldsymbol{E}\cdot\mathbf{d}\boldsymbol{l} = 0 \tag{1.16}$$

The principle of conservation of energy often provides the best way of calculating the velocities of charged particles in electric fields.

Worked Example 1.2

An electron starts with zero velocity from a cathode which is at a potential of -10 kV and then moves into a region of space where the potential is zero. Find its velocity.

Solution Let the velocity of the electron be v and its mass m; then its kinetic energy is $\frac{1}{2}mv^2$. Applying the principle of conservation of energy we obtain $\frac{1}{2}mv^2 = qV_0$, where q is the charge on the electron and V_0 is the cathode potential. Hence $v = (2qV_0/m)^{1/2} = 59.3 \times 10^6\,\mathrm{m\,s^{-1}}$, since $(q/m) = 1.759 \times 10^{11}\,\mathrm{C\,kg^{-1}}$. For accelerating voltages much above 10 kV relativistic effects become important because the electron velocity is comparable with the velocity of

A useful introduction to the theory of relativity is given by Rosser, W.G.V. *An Introduction to the Theory of Relativity*, Butterworths (1971).

light ($2.998 \times 10^8\,\mathrm{m\,s^{-1}}$). It is then necessary to use the correct relativistic expression for the kinetic energy of the electron, but the principle of the calculation is unchanged.

Calculation of potential in simple cases

In simple cases where the electric field can be calculated by using Gauss' theorem it is possible to calculate the potential by using Equation (1.13). More complicated problems can be solved by using the principle of superposition. Since scalar quantities are much easier to add than vectors it is best to superimpose the potentials rather than the fields. The following example shows the method.

Worked Example 1.3

Figure 1.6 shows a cross-sectional view of two long straight cylindrical rods each of radius a. The rods are parallel to each other with their centre lines d apart. Rod A carries a charge q per unit length uniformly distributed and rod B carries a similar charge $-q$. Find an expression for the electrostatic potential at any point on the plane passing through the centre lines of the rods.

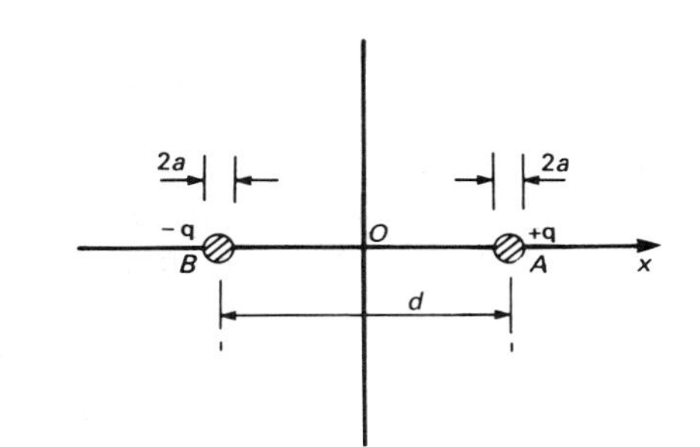

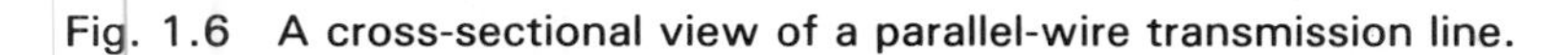

Fig. 1.6 A cross-sectional view of a parallel-wire transmission line.

Solution The electric field of either rod on its own can be found by applying Gauss' theorem as in Worked Example 1.1, with the result given by Equation (1.6). Since $\boldsymbol{E}$ is everywhere radial it follows that V depends only on r and

$$V = -\int \frac{q}{2\pi\epsilon_0 r}\,\mathrm{d}r = -\frac{q}{2\pi\epsilon_0}\ln r + \text{constant} \tag{1.17}$$

where the constant of integration can be given any convenient value because the zero of potential is arbitrary.

It is appropriate to choose the origin of coordinates to be at O, mid-way between the rods, because of the symmetry of the problem, so that OA lies along the x-axis. Then for rod A, $r = |x - \frac{1}{2}d|$. The same argument can be used for rod B, giving an expression for the potential which is identical to (1.17) except that the sign is reversed and $r = |x + \frac{1}{2}d|$. Superimposing these two results and substituting the appropriate expressions for the radii we get:

The modulus sign is needed to ensure that r has the correct value at points on both sides of the rod.

$$V = \frac{q}{2\pi\epsilon_0} \ln \left| \frac{2x + d}{2x - d} \right| \tag{1.18}$$

where the constant of integration has been set equal to zero. This choice makes $V = 0$ when $x = 0$. The same method could be used to find a general expression for the potential at any point in space.

Calculation of the electric field from the potential

We have so far been concerned with means of calculating the potential from the electric field. In many cases it is necessary to reverse the process and calculate the field from a known potential distribution. Figure 1.7 shows a small movement $\mathbf{d}\boldsymbol{l}$ which may be expressed in terms of its components as

$$\mathbf{d}\boldsymbol{l} = \hat{\boldsymbol{x}}\,\mathrm{d}x + \hat{\boldsymbol{y}}\,\mathrm{d}y + \hat{\boldsymbol{z}}\,\mathrm{d}z \tag{1.19}$$

The electric field may likewise be expressed in terms of its components

$$\boldsymbol{E} = \hat{\boldsymbol{x}}\,E_x + \hat{\boldsymbol{y}}\,E_y + \hat{\boldsymbol{z}}\,E_z \tag{1.20}$$

Then from Equations (1.12), (1.19) and (1.20) the potential change along $\mathbf{d}\boldsymbol{l}$ is given by

$$\begin{aligned} \mathrm{d}V &= -(\hat{\boldsymbol{x}}\,E_x + \hat{\boldsymbol{y}}\,E_y + \hat{\boldsymbol{z}}\,E_z)\cdot(\hat{\boldsymbol{x}}\,\mathrm{d}x + \hat{\boldsymbol{y}}\,\mathrm{d}y + \hat{\boldsymbol{z}}\,\mathrm{d}z) \\ &= -(E_x\,\mathrm{d}x + E_y\,\mathrm{d}y + E_z\,\mathrm{d}z) \end{aligned} \tag{1.21}$$

Setting $\mathrm{d}y = \mathrm{d}z = 0$ we have, for a movement in the x-direction,

$$E_x = -\frac{\partial V}{\partial x}$$

with similar expressions for the other two coordinate directions. By superposition the total electric field is

$$\boldsymbol{E} = -\left(\hat{\boldsymbol{x}}\frac{\partial V}{\partial x} + \hat{\boldsymbol{y}}\frac{\partial V}{\partial y} + \hat{\boldsymbol{z}}\frac{\partial V}{\partial z}\right) \tag{1.22}$$

The expression in parentheses on the right-hand side of Equation (1.22) is termed

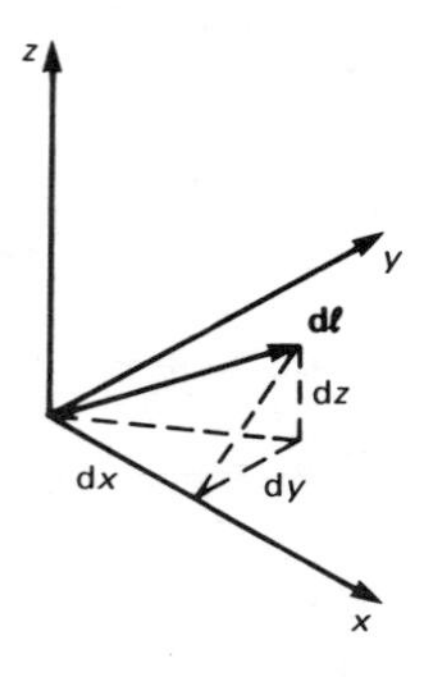

Fig. 1.7 A small vector $\mathbf{d}\boldsymbol{\ell}$ can be regarded as the sum of vectors of magnitude dx, dy and dz along the coordinate directions.

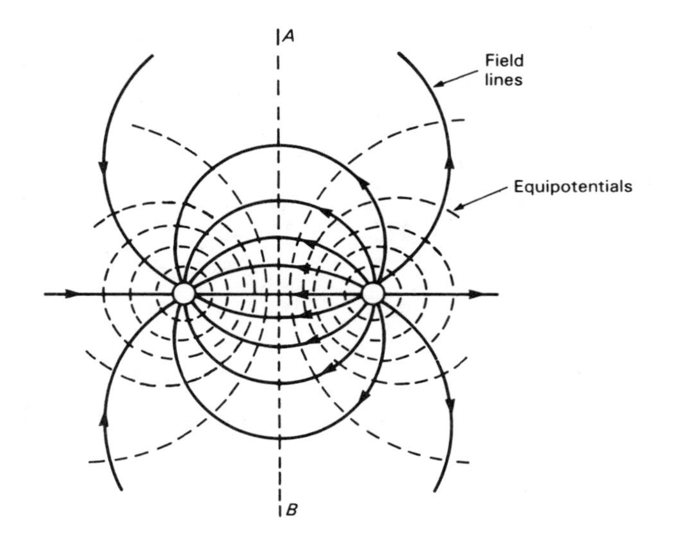

Fig. 1.8 The field pattern around a parallel-wire transmission line.

the **gradient** of V. It can be obtained by operating on V with the operator ∇ defined by Equation (1.10) so that Equation (1.22) may be written

$$\boldsymbol{E} = -\text{grad } V = -\nabla V \tag{1.23}$$

This equation, like Equation (1.11), can be written in terms of any orthogonal coordinate system by using the appropriate form for ∇V.

The gradient of the gravitational potential on a hillside is a vector whose magnitude is equal to the slope of the hillside in the direction of steepest ascent, and whose direction is the direction of steepest ascent. Contour lines pass through points which are all at the same height. They are thus lines of constant gravitational potential and the direction of steepest ascent is always at right angles to them. From Equation (1.12) it can be seen that if $\mathbf{d}\boldsymbol{l}$ lies in such a direction that V is constant it must be perpendicular to $\boldsymbol{E}$. Surfaces on which V is constant are known as **equipotential surfaces** or just **equipotentials**. They always intersect the lines of $\boldsymbol{E}$ at right angles. It has already been mentioned that field plots are useful aids to thought in electrostatics. They can be made even more useful by the addition of the equipotentials. Figure 1.8 shows, as an example, the field plot for Worked Example 1.3.

Conducting materials in electrostatic fields

A conducting material in the present context is one which allows free movement of electric charge within it on a time scale which is short compared with that of the problem. Under this definition metals are always conductors but some other materials which are insulators on a short time scale may allow a redistribution of charge on a longer one. They may be regarded as conducting materials in electrostatic problems if we are prepared to wait for long enough for the charges to reach equilibrium. The charge distribution tends to equilibrium as $\exp(-t/\tau)$, where the time constant t is known as the **relaxation time**. Some typical values are:

The steady flow of electric current is dealt with in Chapter 3.

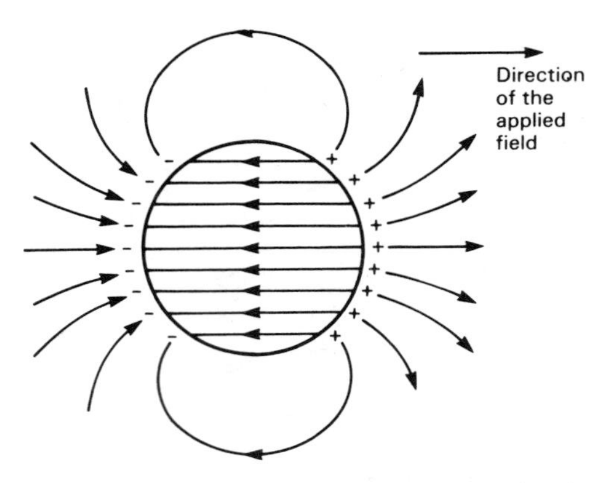

Fig. 1.9 The field pattern of the charge induced on a conducting sphere placed in a uniform electric field.

copper	1.5×10^{-19} s
distilled water	10^{-6} s
fused quartz	10^{6} s

Once the charges have reached equilibrium there can be no force acting on them and the electric field within the material must be zero.

When an uncharged conducting body is placed in an electric field, the free charges within it must redistribute themselves to produce zero net field within the body. Consider, for example, a copper sphere placed in a uniform electric field. The copper has within it about 10^{29} conduction electrons per cubic metre, and their charge is balanced by the equal and opposite charge of the ionic cores fixed in the crystal lattice. The available conduction charge is of the order of $10^{10}\,\mathrm{C\,m^{-3}}$, and only a tiny fraction of this charge has to be redistributed to cancel any practicable electric field. This redistribution gives rise to a surface charge, somewhat as shown in Fig. 1.9, whose field within the sphere is exactly equal and opposite to the field into which the sphere has been placed. This surface charge is known as **induced charge**. It is important to remember that the positive and negative charges balance so that the sphere still carries no net charge. The complete solution to the problem is obtained by superimposing the original uniform field on that shown in Fig. 1.9 to give the field shown in Fig. 1.10. Note that the flux lines must meet the conducting surface at right angles because the surface is an equipotential.

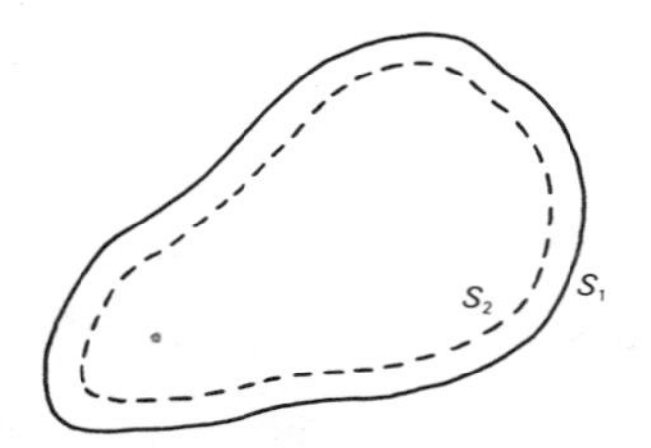

Not only is there no electric field within a conducting body, but there is also no field within a closed conducting shell placed in an electric field. To prove this, consider the figure in the margin, which shows a closed conducting shell S_1. This must be an equipotential surface. If there is any electric field within S_1 there must be other equipotentials such as S_2 lying wholly within S_1. Now the interior of the shell contains no free charge so, applying Gauss' theorem to S_2, the flux of $\boldsymbol{E}$ out of S_2 is zero. But, since it has been postulated that S_2 is an equipotential surface, this can be true only if $\boldsymbol{E}$ is zero everywhere on it and the potential of S_2 is the same as that of S_1. A closed hollow earthed conductor can therefore be used to screen sensitive electronic equipment from electrostatic interference. The screening is perfect as long as there are no holes in the enclosure to allow wires, for

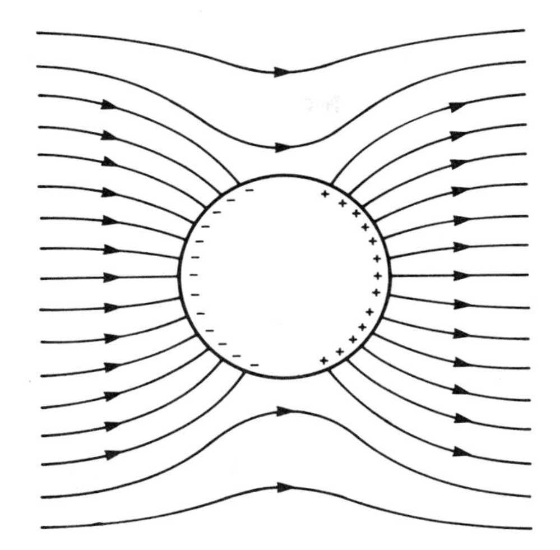

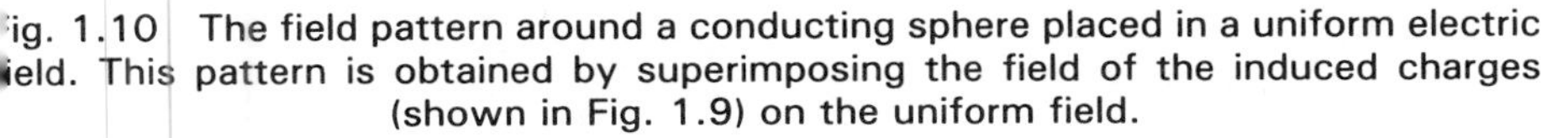

ig. 1.10 The field pattern around a conducting sphere placed in a uniform electric ield. This pattern is obtained by superimposing the field of the induced charges (shown in Fig. 1.9) on the uniform field.

xample, to pass through. Even when there are holes in the enclosure the screening an still be quite effective, for reasons which will be discussed in the next chapter. When the electric field varies with time other screening mechanisms come into play nd the screening is no longer so perfect. Magnetic interference is discussed in Chapters 5 and 6, and radiofrequency interference in Chapter 9.

Methods of solving electrostatic field problems

We have already seen that the electric field produced by a known distribution of harges can be calculated in simple cases, by the application of Gauss' theorem and he principle of superposition. In most practical problems, however, the charge is nknown and the problem is specified in terms of the potentials on electrodes. Simple problems of this type can be solved by the use of Gauss' theorem if it is ossible to make assumptions about the distribution of charges from the symmetry of the problem.

Worked Example 1.4

An air-spaced transmission line consists of two parallel cylindrical conductors ach 2 mm in diameter with their centres 10 mm apart. Assuming that the electrical reakdown strength of air is 3 MV m^{-1}, calculate the maximum potential differnce which can be applied to the conductors.

Solution Since the diameters of the conductors are small compared with their separation it is reasonable to assume that close to the surface of each conductor he field pattern is determined almost entirely by that conductor. The equipotenial surfaces close to the conductors take the form of coaxial cylinders, as may be seen in Fig. 1.8. This is equivalent to assuming that the two conductors can be epresented by uniform line charges $\pm q$ along their axes. The electric field of ither cylinder is then given by Equation (1.6) (for $r \geqslant 1$ mm) with the appropriate

It is not strictly correct to regard this as a 'free-space' problem. In practice air and other gases at atmospheric pressure are indistinguishable from free space until voltage breakdown occurs.

sign for q. Since the strength of the electric field of each line charge is inversely proportional to the distance from the charge, the greatest electric field must occur on the plane passing through the axes of the two conductors. Using the notation of Fig. 1.6 and Equation (1.6) the electric field on the x axis is given by

$$E = \frac{q}{2\pi\epsilon_0(x - d/2)} - \frac{q}{2\pi\epsilon_0(x + d/2)}$$

It is easy to show that this expression is a maximum on the surfaces on the rods (as might be expected from Fig. 1.8), that is, when $x = \pm(d/2 - a)$. The maximum charge is therefore given by

$$q_{max} = 2\pi\epsilon_0 E_{max} (d - a)a/d$$

The maximum potential at A is obtained by substituting this expression into Equation (1.18) to give

$$V_A = E_{max} \frac{a(d - a)}{d} \ln \frac{(d - a)}{a}$$

The potential at B is $-V_A$ so the maximum potential difference between the wires is $2V_A$. Substituting the numbers gives the maximum voltage between the wires as 11.9 kV. When the wires are not thin compared with their separation the method of solution is similar but, as can be seen from the equipotentials in Fig. 1.8, the line charges are no longer located at the centres of the wires.

The method of images

If an uncharged, isolated, conducting sheet is placed in an electric field, then equal positive and negative charges are induced on it. Normally this process requires currents to flow in the plane of the sheet, and the field pattern is changed so that the sheet becomes an equipotential surface. If, however, the sheet is arranged so that it coincides with an equipotential surface, the direction of current flow is normal to the plane of the sheet and the two surfaces become oppositely charged. If the sheet is thin, the separation of the positive and negative charges is small and the field pattern is not affected by the presence of the sheet. This fact can be used to extend the range of problems which can be solved by elementary methods. For example, a conducting sheet can be placed along the equipotential AB in Fig. 1.8. It screens the two charged wires from each other so that either could be removed without affecting the field pattern on the other side of the sheet. Thus the field pattern between a charged wire and a conducting plane is just half of that of a pair of oppositely charged conducting wires.

The field between a charged wire and a conducting plane can be found by reversing the train of thought. We note that an **image charge** can be placed on the opposite side of the plane to produce a field which is the mirror image of the original field. The image charge is equal in magnitude to the original charge, but has the opposite sign. The plane is an equipotential surface in the field of the two charges, so it can be removed without altering the field pattern. The problem is then reduced to the superposition of the fields of the original and image charges. This method is known as the **method of images**. It can be applied to the solution of any problem involving charges and conducting surfaces if a set of image charges can be found such that the equipotentials of the whole set of charges in free space coincide with the conducting boundaries.

Worked Example 1.5

A wire 1 mm in diameter is placed mid-way between two parallel conducting planes 10 mm apart. Given that the planes are earthed and the wire is at a potential of 100 V, find a set of image charges that will enable the electric field pattern to be calculated.

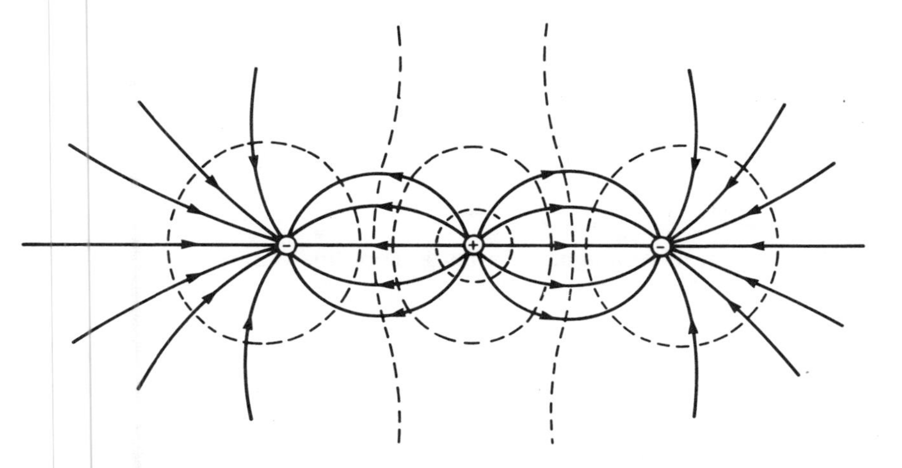

Fig. 1.11 The field pattern around a positively charged wire flanked by a pair of negatively charged wires.

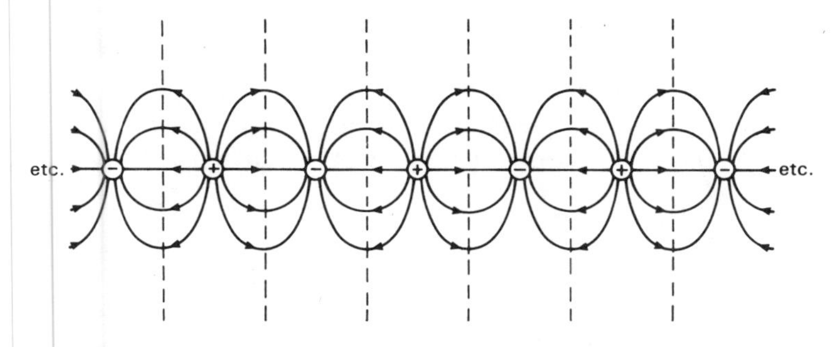

Fig. 1.12 The field pattern around a set of equispaced parallel wires charged alternately positive and negative.

Solution If we were to put just one image charge on either side of the wire the field pattern would be as shown in Fig. 1.11. None of the equipotential surfaces is a plane. The solution is to use an infinite set of equispaced wires charged alternately positive and negative, as shown in Fig. 1.12. The symmetry of this set of wires is such that there must be equipotential planes mid-way between the wires.

Useful collections of the solutions to electrostatic problems can be found in Smythe* and Binns and Lawrenson.*

Laplace's and Poisson's equations

The methods described in the previous section have been applied with ingenuity to a wide variety of problems whose solutions can be looked up when required. Unfortunately engineers are not free to choose the problems they wish to solve, and the great majority of practical problems cannot be solved by elementary methods. Figure 1.13 shows a typical problem: an electron gun. In this case the field problem and the equations of motion of the electrons must be solved simultaneously because the space charge of the electrons affects the field solution. A general method which can be used, in principle, to solve any problem is obtained by combining Equations (1.9) and (1.22):

$$\frac{\partial^2 V}{\partial x^2} + \frac{\partial^2 V}{\partial y^2} + \frac{\partial^2 V}{\partial z^2} = -\rho/\epsilon_0 \tag{1.24}$$

This is known as **Poisson's equation**. It can also be written

$$\nabla^2 V = -\rho/\epsilon_0 \tag{1.25}$$

where ∇^2 is given, in rectangular Cartesian coordinates, by

$$\nabla^2 = \frac{\partial^2}{\partial x^2} + \frac{\partial^2}{\partial y^2} + \frac{\partial^2}{\partial z^2} \tag{1.26}$$

When there is no free charge present the equation takes the simpler form known as **Laplace's equation**:

$$\nabla^2 V = \frac{\partial^2 V}{\partial x^2} + \frac{\partial^2 V}{\partial y^2} + \frac{\partial^2 V}{\partial z^2} = 0 \tag{1.27}$$

See Binns and Lawrenson* for the details of this method and examples of what can be done with it.

This equation has been solved for a very wide range of boundary conditions by analytical methods employing a variety of coordinate systems and by the special method known as **conformal mapping**, which applies to two-dimensional problems. These solutions can be looked up when they are required. Cases whose solutions are not available in the literature must, in nearly every case, be solved by numerical methods. When free charges are present in a problem it is necessary

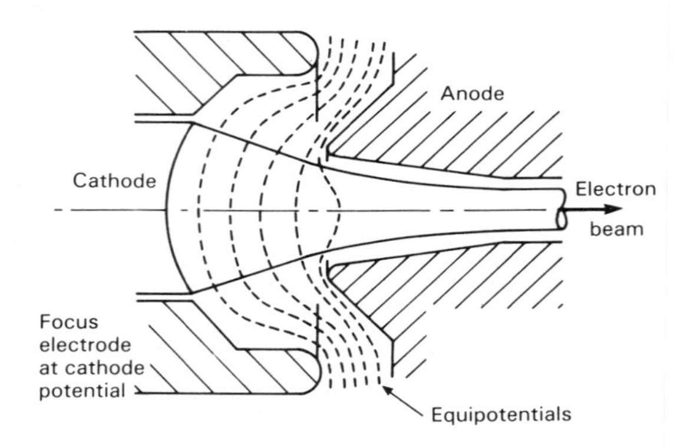

Fig. 1.13 The arrangement of a typical high-power electron gun. Such a gun might produce a 3 A electron beam 10 mm in diameter for a potential difference between cathode and anode of 20 kV. Electron guns like this one are used in klystrons for producing the output power in television transmitters.

to use Poisson's equation as the basis of either an analytical or a numerical solution. There are only a few cases which can be solved analytically. One of these is the space-charge limited vacuum diode which is included as Problem 1.8 at the end of this chapter. Another is the depletion layer in a p–n junction diode, but that involves ideas about the electric field in material media, so it must be deferred until those have been considered in Chapter 2.

The finite difference method

The simplest numerical method for solving field problems is the **finite difference method**. In this method a regular rectangular mesh is superimposed upon the problem. The real continuous variation of potential with position is then approximated by the values of the potential at the intersections of the mesh lines. Figure 1.14 shows a small section of a two-dimensional mesh with a spacing h in each direction. To find an approximate relationship between the potentials shown we apply Gauss' theorem to the surface shown by the broken line. The component of the electric field normal to the section AB of the surface is given approximately by

$$E_{nAB} = (V_0 - V_2)/h \tag{1.28}$$

The flux of $\boldsymbol{E}$ through unit length of the face AB is therefore

$$\begin{aligned}\Phi_{AB} &= E_{nAB}h \\ &= V_0 - V_2\end{aligned} \tag{1.29}$$

If the Gaussian surface does not enclose any charge the net flux of E out of it must be zero so that

$$4V_0 - V_1 - V_2 - V_3 - V_4 = 0$$

$$\text{or} \quad V_0 = \tfrac{1}{4}(V_1 + V_2 + V_3 + V_4) \tag{1.30}$$

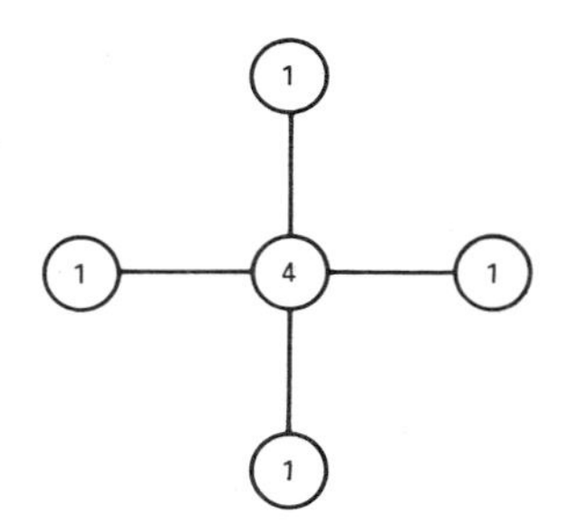

Thus, if we know the potentials at points 1 to 4 approximately, we can use Equation (1.30) to obtain an estimate of V_0. Because the errors in the four potentials cancel each other out to some extent, and because the resulting error is divided by 4, the error in the value of V_0 is normally less than the errors in the potentials used to calculate it. Equation (1.30) is conveniently summarized by the

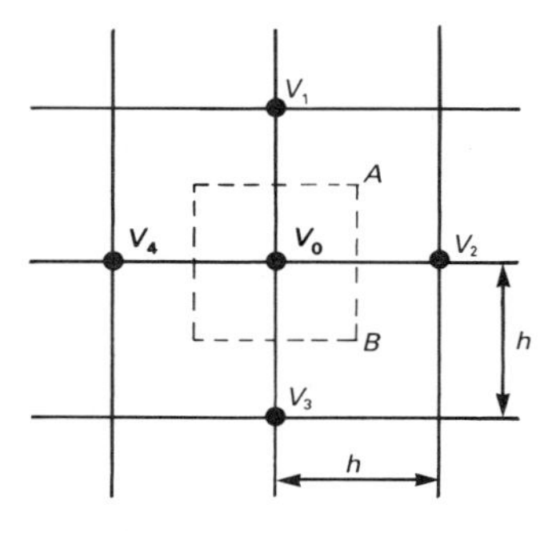

Fig. 1.14 Basis of the finite difference calculation of potential.

diagram in the margin. The use of the method is illustrated by the worked example which follows.

Worked Example 1.6

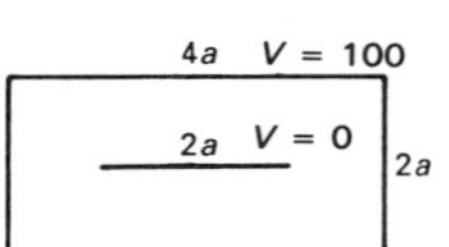

Find the potential distribution between a long thin conducting strip and a surrounding rectangular conducting tube, as shown in the figure in the margin, when the potential difference between them is 100 V.

Solution The problem may be simplified by observing that the solution is the same in each quadrant, subject to appropriate reflections about the planes of symmetry. One quadrant of the diagram is redrawn on an enlarged scale in Fig. 1.15(a) with a square mesh added to it.

To start the solution we first write down the potentials on the electrodes and estimate them at all the interior mesh points. An easy way to do this is to assume that the potential varies linearly with position. These potentials are written along-

(a)

P								
100	100	100	100	100	100	100	100	100
~~75~~ 75	~~75~~ 75	~~75~~ 75	~~75~~ 75	~~75~~ 75	~~75~~ 76	~~75~~ 78	~~75~~ 88	A 100
~~50~~ 50	~~50~~ 50	~~50~~ 50	~~50~~ 50	~~50~~ 51	~~50~~ 54	~~50~~ 64	~~75~~ 78	100
~~25~~ 25	~~25~~ 25	~~25~~ 25	~~25~~ 26	~~25~~ 29	~~25~~ 40	~~50~~ 54	~~75~~ 76	100
0 (Q)	0	0	0	0 (R)	~~25~~ 33	~~50~~ 52	~~75~~ 76	100 S

(b)

100	100	100	100	100	100	100	100	100
75.9	76.1	76.9	78.6	81.6	86.1	91.0	95.6	100
51.3	51.6	52.9	55.8	61.7	71.9	82.3	91.6	100
25.9	26.2	27.3	30.1	37.5	57.6	74.7	88.3	100
0	0	0	0	0	46.5	70.7	86.8	100

Fig. 1.15 The finite difference solution for one quadrant of the problem discussed in Worked Example 1.6: (a) the initial stages and (b) the final solution.

side the mesh points as shown. Next we choose a starting point such as A and work through the mesh, generating new values of the potentials with Equation (1.30). As each new value is calculated it is written down and the previous estimate crossed out. Figure 1.15(a) shows the results of the first pass through the mesh working along each row from right to left. Along the lines PQ and RS we make use of the symmetry of the field to supply the potentials at the mesh points outside the figure (i.e. $V_4 = V_2$ on PQ and $V_3 = V_1$ on RS). Check the figures for yourself and carry the process on for one more pass through the mesh to see how the solution develops. It is not necessary to retain many significant figures in the early stages of the calculation because any errors introduced do not stop the method from converging. If we work to two significant figures we can avoid the use of decimal points by choosing the electrode potentials at 0 and 100 V. The final values of the potentials can be scaled to any other potential difference if required.

The process is carried on until the changes in the potentials on successive iterations fall below some preset limit. A large number of iterations may be required before the solution is complete so it is better to program a computer to do the calculations than to do them by hand. Nevertheless, it is instructive to try the hand calculation at least once in order to understand how the process works. Computer packages are available for the solution of field problems but, if you do not have access to one, it is quite easy to solve problems like this one using a spreadsheet on a personal computer. Figure 1.15(b) shows the solution obtained in this way after about 100 iterations.

It must be emphasized that the solution obtained by this method is approximate because of the discretization of the problem. Improved accuracy can be obtained by using a finer mesh.

From this solution we can obtain the field components at any mesh point. We can also plot the equipotential curves by interpolation between the potentials. The use of the solution to calculate the capacitance per unit length between the electrodes is discussed in Chapter 2.

The method can be applied to problems such as that shown in Fig. 1.13 in which the electrodes are three-dimensional and do not fit a rectangular mesh. Details of the ways in which such problems can be solved can be found in the literature.

Details of finite difference methods and their use can be found in Binns and Lawrenson*

The motion of charged particles in electric fields

In every kind of active electronic device electric fields are used to control the motion of charged particles. Examples are the use of electrostatic lenses to focus electron beams in television picture tubes, electrostatic deflection systems in tubes for laboratory oscilloscopes and electron microscopes, and the control of the current crossing the emitter junction of a bipolar transistor. When the charge densities are small it is possible to calculate the electrostatic fields, neglecting the contributions of the charges to them, and then to integrate the equations of motion of the particles. At higher charge densities the fields are affected by the space charge and it is necessary to find mutually consistent solutions of Poisson's equation and the equations of motion.

Worked Example 1.7

Figure 1.16 shows a simplified form for the deflection plates of a cathode ray tube. Given that the electron beam is launched from an electrode (the cathode)

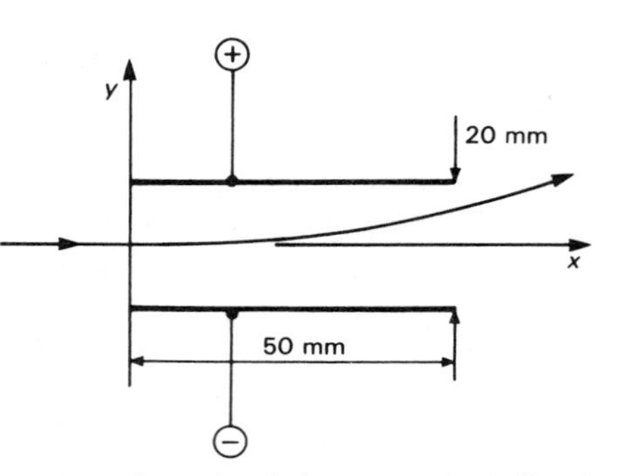

Fig. 1.16 The arrangement of a pair of electrostatic deflection plates for a cathode ray tube.

at a potential of −2000 V and passes between the deflection plates as shown, estimate the angular deflection of the beam when the potentials of the plates are ±50 V.

Solution To make the problem easier we assume that the electric field is constant everywhere between the plates and falls abruptly to zero at the ends. Then the field between the plates is $E_y = -5000 \text{ V m}^{-1}$.

The equation of motion in the y direction for an electron is

$$m\frac{d^2y}{dt^2} = -qE_y \tag{1.31}$$

where q is the magnitude of the electronic charge. The y-component of the velocity of the electrons as they leave the plates is obtained by integrating Equation (1.31) over the time taken for them to pass along the length of the plates. Thus

$$v_y = -(q/m)E_y(0.05/v_x) \tag{1.32}$$

Now v_x can be obtained by using the principle of conservation of energy as in Worked Example 1.2, with the result $v_x = 26.5 \times 10^6 \text{ m s}^{-1}$. Substitution into Equation (1.32) gives $v_y = 1.66 \times 10^6 \text{ m s}^{-1}$, and the angular deflection experienced by the electrons is then arctan $(v_y/v_x) = 3.58°$.

It is, of course, unrealistic to assume that the field between the plates has the idealized form chosen above. To obtain a more accurate estimate of the deflection it would be necessary to find the field distribution between the plates by solving Laplace's equation. Equation (1.31) could then be integrated using a more realistic expression for E_y. Similar calculations can be performed for the motion of the electrons through the electrostatic lenses used to focus them down to a small spot on the screen of a cathode ray tube.

For further information about electrostatic lenses see Spangenburg.*

Summary

In this chapter, starting from the inverse square law of force between two charges, we have derived a range of methods for solving practical problems involving electric fields in free space and the motion of charged particles in them. The concepts of electric field, flux density and potential have been shown to be useful for these purposes.

The very limited range of problems which can be solved by elementary methods can be extended by the use of the principle of superposition and the method of images. In most real problems, however, the electric field can be calculated only by solving Laplace's or Poisson's equations. Reference has been made to books in which the solutions to many such problems can be looked up when they are needed. Cases which have not been solved before generally have to be tackled using numerical methods, as illustrated by a worked example.

Finally, the motion of charged particles in electric fields has been considered. The ideas contained in this chapter find their direct application in problems about voltage breakdown between electrodes and those dealing with the motion of charged particles. They can also be applied to the calculation of capacitance, as we shall see in the next chapter.

Problems

1.1 The surface charge density on a metal electrode is $\sigma \,\mathrm{C\,m^{-2}}$. Use Gauss' theorem to show that the electric field strength close to the surface is $E = \sigma/\epsilon_0$.

1.2 An electron beam originating from a cathode at a potential of -10 kV has a current of 1 A and a radius of 10 mm. The beam passes along the axis of an earthed conducting cylinder of radius 20 mm. Use Gauss' theorem to find expressions for the radial electric field within the cylinder, and calculate the potential on the axis of the system.

What effect does this have on the velocity of the electrons? Use revised estimates of the velocities of the electrons to produce a revised figure for the space-charge potential. Repeat the process until mutually consistent values have been found.

1.3 Calculate the radial force on an electron on the surface of the electron beam in the previous question. Estimate the distance the beam could be projected along the cylinder before electrons began to be intercepted by it if their initial motion is parallel to the axis.

It is important to realize that in arriving at the apparently exact solutions to problems like 1.4 we have nearly always made some approximations. Most coaxial lines have braided outer conductors, so the approximation is to replace that by a smooth continuous conducting cylinder.

1.4 An air-spaced coaxial line has inner and outer conductors with radii R_1 and R_2 respectively. Show that the breakdown voltage of the line is highest when $\ln(R_2/R_1) = 1$.

1.5 A metal sphere of radius 10 mm is placed with its centre 0.5 m from a flat earthed sheet of metal. Assuming that the breakdown strength of air is $3\,\mathrm{MV\,m^{-1}}$, calculate the maximum voltage which can be applied to the electrode without breakdown occurring. What is then the surface-charge density on the sphere?

1.6 The figure in the margin shows a charged wire which is equidistant from a pair of earthed conducting planes which are at right angles to each other. Where should image charges be placed in order to solve this problem by the method of images? What difference would it make if the planes were at 60°

+
d
d

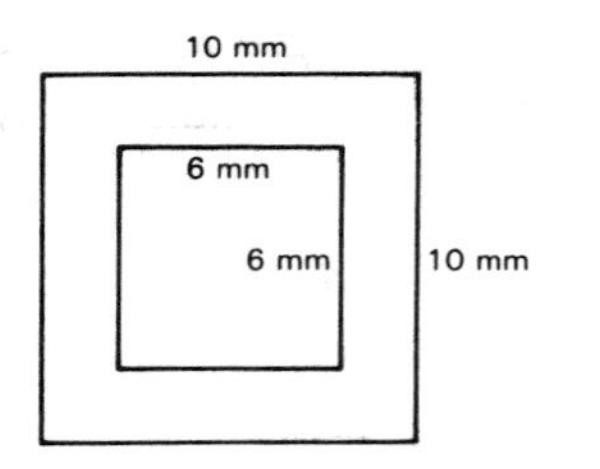

When a diode is operated under these conditions the current is said to be **space-charge limited.** The current density at the surface of a cathode is very strongly dependent upon the temperature of the cathode and the condition of its surface. By using space-charge limitation we can ensure that the properties of diodes (used mostly in electron guns nowadays) depend only on their dimensions.

to each other? Could the method be used when the planes were at 50° to each other? How could the solution be deduced for any angle between 0° and 90°?

1.7 The figure in the margin shows a square coaxial arrangement of electrodes. If the potential of the inner electrode is 5 V above that of the outer electrode estimate the maximum and minimum values of the electric field in the space between the electrodes.

1.8 A simple thermionic diode consists of two plane parallel electrodes: the cathode and the anode. Electrons are emitted from the surface of the cathode with zero velocity and accelerated towards the anode which is maintained at a potential V_a with respect to the cathode. If the density of electrons between the electrodes is great enough the space charge alters the distribution of the electric field. Show that, in the limit of high space-charge density, the current through the diode is proportional to $V_a^{3/2}$ and independent of the rate at which electrons are supplied by the cathode.

Dielectric materials and capacitance

2

Objectives

- ☐ To discuss how and why an electric field is affected by the presence of dielectric materials.
- ☐ To introduce the electric flux density vector $\boldsymbol{D}$ as a useful tool for solving problems involving dielectric materials.
- ☐ To derive the boundary conditions which apply at the interface between different dielectric materials.
- ☐ To introduce the idea of capacitance as a general phenomenon which is not restricted to capacitors.
- ☐ To demonstrate the calculation of capacitance by the use of Gauss' theorem, field solutions and energy methods.
- ☐ To introduce the idea of stored energy density in an electric field.
- ☐ To discuss the causes of electrostatic interference and techniques for reducing it.

Insulating materials in electric fields

Very many materials do not allow electric charges to move freely within them, or allow such motion to occur only very slowly. These materials are not only used to block the flow of electric current but also to form the insulating layer between the electrodes of capacitors. In this context they are known as **dielectric** materials. By making an appropriate choice of dielectric material for a capacitor it is possible to reduce the size of a capacitor of given capacitance or to increase its working voltage. If a dielectric material is subjected to a high enough electric field it becomes a conductor of electricity, undergoing what is known as **dielectric breakdown**. This controls the maximum working voltage of capacitors, the maximum power which can be handled by coaxial cables in high-power applications such as radio transmitters, and the maximum voltages which can be sustained in microcircuits. It is not always appreciated that because dielectric breakdown depends on the electric field strength it can occur when low voltages are applied across very thin pieces of dielectric material.

Examples of insulating materials are gases, organic liquids, rubber, plastics, glasses and ceramics.

The gate insulation of an MOS transistor may break down with only 30 V across it.

In order to understand the behaviour of dielectric materials in electric fields it is helpful to make a comparison with that of conductors. Figures 2.1(a) and (b) show respectively a conducting sheet and a dielectric sheet placed between parallel electrodes to which a potential difference has been applied. The potential difference is associated with equal and opposite charges on the two electrodes. The conducting sheet of Fig. 2.1(a) contains electrons which are free to move and set up a surface charge which exactly cancels the electric field within the conductor in the manner discussed in Chapter 1. The electrons in the dielectric material, on the other hand, are bound to their parent atoms and can only be displaced to a

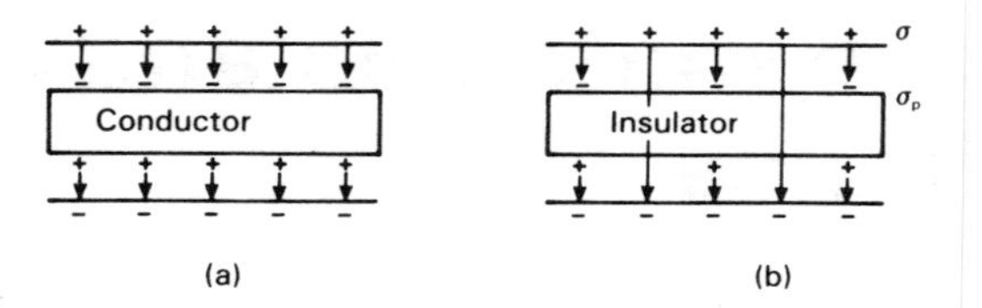

Fig. 2.1 A comparison between a conductor and an insulator placed in an electric field. The conductor (a) acquires a surface charge whose field exactly cancels the external field within the conductor. The insulator (b) is polarized, but the surface charges only reduce the field within it; they are not large enough to cancel it.

limited extent by the applied electric field. This displacement, however, is sufficient to produce some surface charge and the dielectric is then said to be **polarized**. The surface charge is not sufficient to cancel the electric field within the sheet, but it does reduce it to some extent, as shown in Fig. 2.1(b). Polarization may also produce a volume distribution of charge, but we shall assume that this does not occur in the materials in which we are interested. It is important to remember that the surface charge produced by the polarization of a dielectric is a bound charge which, unlike the surface charge induced on a conductor, cannot be removed. The polarization charges must also be carefully distinguished from any free charge which may reside on the surface of a dielectric.

On materials which are good insulators, free charges may persist for long periods and strong electric fields may build up as a result of them. These phenomena have many important practical consequences, but they are not easy to study theoretically because the distribution of the charges is usually unknown. In metal oxide semiconductor (MOS) integrated circuits, for example, it is possible for charges to build up on the gate electrodes if they are left unconnected. The electric field produced by these charges can be strong enough to cause dielectric breakdown of the silicon dioxide layer. This is why special precautions have to be taken when handling these circuits. In what follows we shall assume that the dielectric is initially uncharged and that any surface charge is the result of polarization.

To put this subject on a quantitative basis, let us suppose that the electrodes in Fig. 2.1(b) carry a surface charge σ per unit area and that the surface charge on the dielectric is σ_p per unit area. Now, assuming that the electric field is everywhere uniform and normal to the electrodes, the field outside the dielectric is given by

$$E_a = \sigma/\epsilon_0 \tag{2.1}$$

using the result of Problem 1.1, whereas that within the dielectric is

$$E_d = (\sigma - \sigma_p)/\epsilon_0 \tag{2.2}$$

Equation (2.2) can be rewritten as

$$E_d = \frac{\sigma}{\epsilon_0\sigma/(\sigma - \sigma_p)} = \sigma/\epsilon_0\epsilon_r \tag{2.3}$$

In older books ϵ_r is sometimes called the **dielectric constant**.

where $\epsilon_r = \sigma/(\sigma - \sigma_p)$ is known as the **relative permittivity** of the material. Since $\sigma_p < \sigma$ it follows that $\epsilon_r > 1$. It is unfortunate that the symbol ϵ_r has been adopted for this property of dielectric materials because there could be some confusion between it (a dimensionless quantity) and the **permittivity**, defined as $\epsilon = \epsilon_0 \epsilon_r$ and measured in farads per metre. Care must be taken not to get these symbols confused with each other.

In order to make the theory simpler, we shall assume that σ_p is proportional to σ and that ϵ_r is therefore a constant. This assumption holds good for many of the materials used in electronic engineering, but it is very important to remember that it is not always valid. In particular, ϵ_r for some materials may depend on:

- the strength of the electric field;
- frequency (if the field is varying with time);
- the orientation of crystal axes to the field;
- the previous history of the material.

Problems involving linear dielectric materials could be solved by calculating the polarization charges and finding the fields resulting from both the free charges on the electrodes and the bound polarization charges. This would not usually be easy and it is much better to use an approach in which the polarization charges are implicit. To do this we introduce a new vector known as the **electric flux density**, which is defined by

There are other ways of introducing the vector ***D***. See Compton* for one possibility. A more rigorous discussion of the polarization of dielectric materials is given by Bleaney and Bleaney.*

$$\boldsymbol{D} = \epsilon \boldsymbol{E} \tag{2.4}$$

In the example given above the electric flux density outside the dielectric is $D = \sigma$ and what within the dielectric is likewise σ. In other words $\boldsymbol{D}$ depends only on the free charges, unlike $\boldsymbol{E}$, which depends on the polarization charges as well. It can be shown that, subject to the validity of the assumption that ϵ_r is a constant, the argument given above can be generalized to cover pieces of dielectric material of any shape. Gauss' theorem (Equation (1.5)) can thus be written in a form which is valid for problems involving dielectric materials:

This form of Gauss' theorem is always true. It takes the form given in Equation (1.5) when there are no polarization charges present.

$$\oiint \boldsymbol{D} \cdot \mathbf{d}\boldsymbol{S} = \iiint \rho \,\mathrm{d}v \tag{2.5}$$

or, in differential form

$$\nabla \cdot \boldsymbol{D} = \rho \tag{2.6}$$

Solution of problems involving dielectric materials

Many problems in electrostatics deal with sets of electrodes together with dielectric materials. When the symmetry of a problem is simple it is possible to use Gauss' theorem in much the same way as in Chapter 1.

Worked Example 2.1

Figure 2.2 shows a coaxial cable in which the space between the conductors is filled with a dielectric material of permittivity ϵ. Find an expression for the electric field within the dielectric when the potential difference between the electrodes is V_0.

Solution Assume that the conductors carry charges $\pm q$ per unit length, with the inner conductor being positively charged. Applying Gauss' theorem as given in Equation (2.5), we have

$$D = q/2\pi r$$

then, using Equation (2.4),

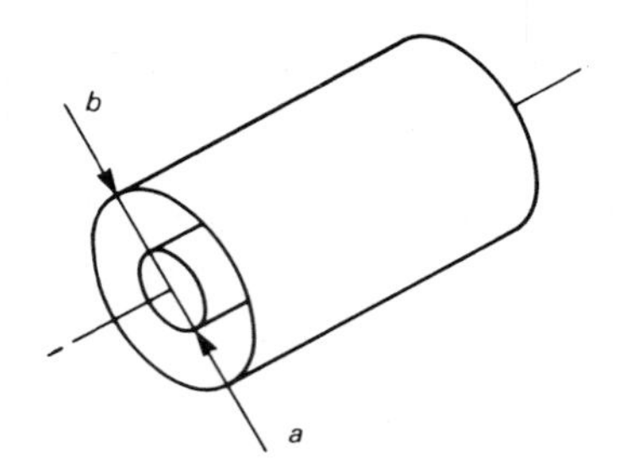

Fig. 2.2 The arrangement of coaxial cylindrical electrodes, an idealization of a coaxial cable.

$$E = D/\epsilon = q/2\pi\epsilon r$$

The potential difference between the conductors is found by using Equation (1.13):

$$V_b - V_a = -\int_a^b E\,\mathrm{d}r = -\int_a^b (q/2\pi\epsilon r)\,\mathrm{d}r$$
$$= -V_0$$

Then

$$V_0 = (q/2\pi\epsilon)\ln(b/a)$$

The charge q which was assumed for the purposes of the solution need not be calculated explicitly and can be eliminated to give the required expression for the electric field:

$$E = V_0/[r\ln(b/a)]$$

Boundary conditions

When two or more dielectric materials are present it is necessary to treat each region separately and then to apply the appropriate *boundary conditions* at the interfaces. There are three of these conditions relating to V, E and D respectively.

If the potential changes abruptly then $\mathrm{d}V/\mathrm{d}x \to \infty$.

The electrostatic potential must always be continuous at a boundary, that is, it cannot change suddenly there. The physical reason for this condition is that an abrupt change in the potential would imply the presence of an infinitely strong electric field.

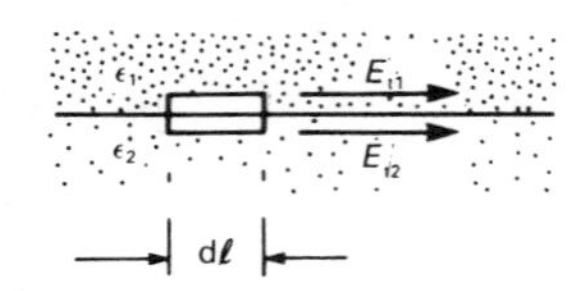

The negative sign of E_{t2} occurs because the path of integration is traversed in a clockwise direction and the direction of integration in material 2 is opposite to that of E_{t2}.

To find the boundary condition for the electric field we consider an infinitesimal closed path as shown in the figure in the margin. The path is chosen so that it crosses the boundary between two dielectrics having permittivities ϵ_1 and ϵ_2 as shown. If the loop is made very thin, then the contributions to the line integral in Equation (1.16) arising from the parts of the loop normal to the boundary are negligible. If, in addition, the tangential components of the electric field are E_{t1} and E_{t2}, then the integral becomes

$$(E_{t1} - E_{t2})\,\mathrm{d}l = 0$$

or

$$E_{t1} = E_{t2} \tag{2.7}$$

This result can be stated in words as: *the tangential component of the electric field is continuous at a boundary.*

The boundary condition for the electric flux density can be found in a similar way by using Gauss' theorem. The figure in the margin shows a boundary between two dielectric materials with an infinitesimal Gaussian surface which crosses it. If the thickness of the 'pill box' is very small, then the contributions to the flux from the parts of the surface which are normal to the boundary are negligible and the integral becomes

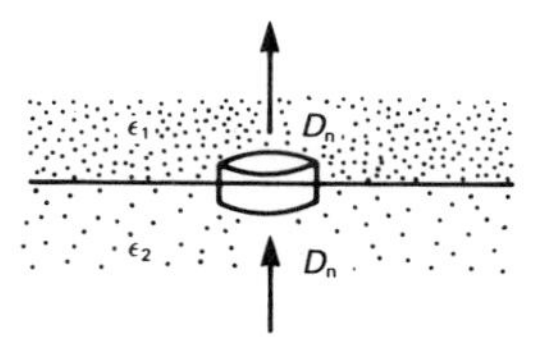

$$(D_{n1} - D_{n2})\, dA = 0$$

where D_{n1} and D_{n2} are the components of $\boldsymbol{D}$ normal to the boundary on each side of it and dA is the area of the part of the boundary lying within the Gaussian surface. Thus

We are assuming here that there are no free charges on the boundary.

$$D_{n1} = D_{n2}$$

or, in words, *the normal component of* **D** *is continuous at a boundary.*

Worked Example 2.2

Figure 2.3 shows a coaxial cable with two layers of dielectric material. Find an expression for the electric field at any point within the dielectric.

Solution As in the simpler case with a single dielectric, we assume a charge q per unit length on the inner conductor. Applying Gauss' theorem gives

Refer back to Worked Example 2.1 for the method of applying Gauss' theorem to this problem. The same geometries are used repeatedly in this book, so that you can see how the solutions to the more complicated problems can be built up from those of simpler ones.

$$D = q/2\pi r$$

as before. From this point on we must consider the two dielectric regions separately. If the radial components of the electric field are E_1 and E_2 in the inner and the outer region, respectively, then

$$E_1 = q/2\pi\epsilon_1 r \qquad \text{and} \qquad E_2 = q/2\pi\epsilon_2 r$$

The potential difference across the inner layer is

$$\begin{aligned} V_b - V_a &= -\int_a^b (q/2\pi\epsilon_1 r)\, dr \\ &= (q/2\pi\epsilon_1) \ln (a/b) \end{aligned}$$

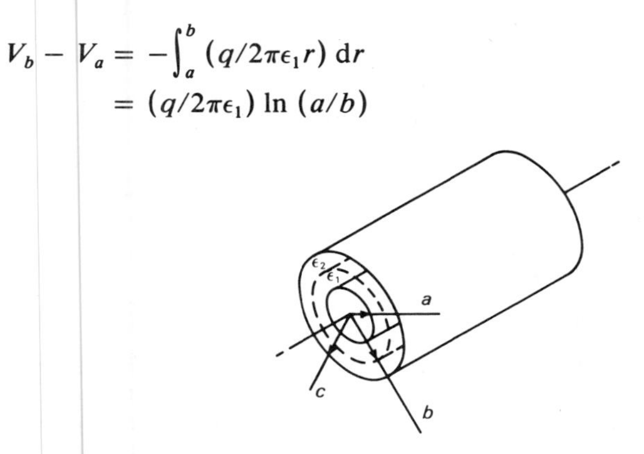

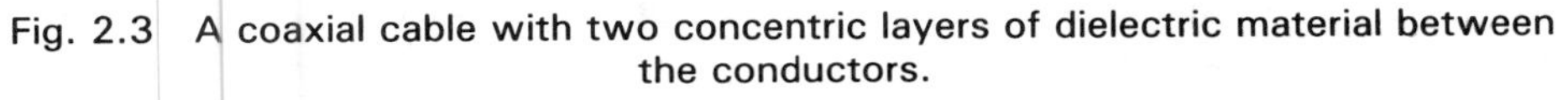
Fig. 2.3 A coaxial cable with two concentric layers of dielectric material between the conductors.

In the same way the potential difference across the outer region is

$$V_c - V_b' = (q/2\pi\epsilon_2)\ln(b/c)$$

Applying the boundary condition for the potential at the interface between the dielectrics, we obtain $V_b = V_b'$. The overall potential difference is thus

$$V_0 = (V_a - V_b) + (V_b - V_c)$$

Substituting for the expressions in the brackets

$$V_0 = \frac{q}{2\pi}\left(\frac{1}{\epsilon_1}\ln(b/a) + \frac{1}{\epsilon_2}\ln(c/b)\right)$$

The expressions for the electric field can then be found as before by eliminating q. Although the other two boundary conditions have not been involved explicitly on this occasion, it is easy to show that they are indeed satisfied by this solution.

The symmetry of the problem ensures that the electric field is entirely radial, so the tangential field is zero on each side of the boundary. The condition on the radial component of the electric flux density is satisfied automatically in the application of Gauss' theorem.

Capacitance

Capacitors are very familiar as circuit elements, but it is not always realized that the idea of capacitance is more general. Capacitance exists between any pair of conductors which are electrically insulated from each other. Thus there is a capacitance between adjacent tracks on a printed circuit board, but this does not appear in the circuit diagram. This 'stray' or 'parasitic' capacitance can cause unwanted coupling between the parts of a circuit, causing it to oscillate or misbehave in some other way. Very few electronic engineers ever need to calculate the capacitance of a capacitor; they are much more likely to need to estimate the magnitude of a stray capacitance.

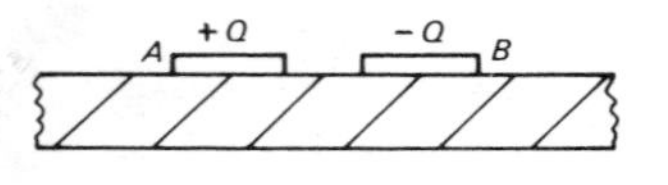

The figure in the margin shows a section through two adjacent tracks on a printed circuit board. Let us suppose that the tracks are insulated from each other and from earth and that charge Q is transferred from B to A. Electrode B must then carry charge $-Q$. As a result of this transfer of charge, an electric field exists around the electrodes such that the potential difference between them is V. If the dielectric material of the printed circuit board has a permittivity which does not vary with the electric field, then the system is linear and the principle of superposition may be applied. It follows that the potential difference between the electrodes is directly proportional to the charges on them, so we can write

$$Q = CV$$

$1\ \mu F = 10^{-6}$ F, 1 nF $= 10^{-9}$ F, 1 pF $= 10^{-12}$ F. Engineers usually use only powers of 10 which are multiples of 3.

where C is a constant of proportionality which is readily recognized as the capacitance between the electrodes familiar from elementary circuit theory. The unit of capacitance is the **farad** (F) and $1\ F = 1\ C\ V^{-1}$. Most capacitances are small and measured in microfarads, nanofarads or picofarads. Stray capacitances are usually of the order of picofarads.

Electrostatic screening

It has already been noted that unwanted capacitive coupling between electronic circuits can be a major problem. This is part of the larger problem of **electromagnetic interference**; another aspect, inductive coupling, is discussed in Chapter 6, and radiofrequency interference in Chapter 9. The problem with all types of electromagnetic interference is how to minimize it rather than how to calculate its magnitude accurately. Electromagnetic theory provides the means for understanding the causes of electromagnetic interference and the techniques for dealing with them.

A dramatic example of electromagnetic interference is the electromagnetic pulse (emp) produced by the explosion of an atomic bomb. This pulse is strong enough to cause permanent damage to unprotected electronic equipment.

For further information about screening against electromagnetic interference consult Morrison* and Sangwine.*

A simple case of the coupling of two circuits by stray capacitance is shown in Fig. 2.4. The circuits 1 and 2 are linked by the stray capacitance C_s and by a common earth. The stray capacitance is small, typically of the order of a picofarad, so its impedance ($Z_s = 1/j\omega C_s$) is high, but decreases with increasing frequency. In this problem V_1 is the source of the interference picked up by circuit 2. The current flowing through the capacitor is small compared with that in R_{L1}, so the spurious signal appearing at the input of the amplifier is approximately

$$V_s = \left\{\frac{R_{L1}}{R_{s1} + R_{L1}}\right\}\left\{\frac{R'_{in}}{Z_s + R'_{in}}\right\} V_1 \tag{2.8}$$

where $Z_s = 1/j\omega C_s$ and $R'_{in} = (R_{s2}R_{in})/(R_{s2} + R_{in})$. The spurious voltage given by Equation (2.8) is to be compared with the signal voltage at the input of the amplifier

$$V_{sig} = V_2 R_{in}/(R_{s2} + R_{in})$$

Equation (2.8) shows that the spurious signal is greatest when the source impedance R_{s1} is low and the effective input impedance of the second circuit R'_{in} is high. An indication of the order of magnitude of capacitance which can cause trouble can be obtained by supposing that circuit 1 is the a.c. main. R_{s1} is then very small. If R'_{in} is 1 MΩ and V_s is 1 μV, Z_s is approximately 2.4×10^{14} Ω, which corresponds at 50 Hz to a stray capacitance of the order of 10^{-17} F.

The capacitive coupling between the circuits can be reduced by putting an earthed screen between them, as shown in Fig. 2.5. The stray capacitance is divided into two parts in series with each other with their common point earthed. In practice, unless the screen completely encloses one of the circuits, there is still a residual capacitance connected directly between P and Q, bypassing the screen. Quite effective screening can be achieved with a partial enclosure provided that it intercepts most of the field lines passing from P to Q. It is essential that the screen

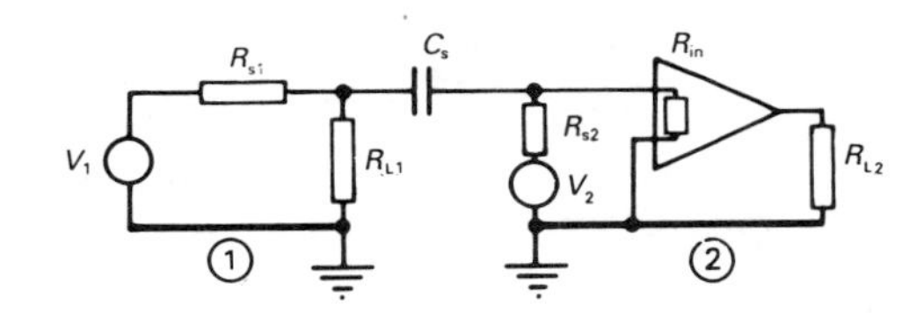

Fig. 2.4 Showing how two circuits can be coupled together by a stray capacitance C_s. The unwanted signal coupled from circuit 1 into circuit 2 can be large enough to be troublesome.

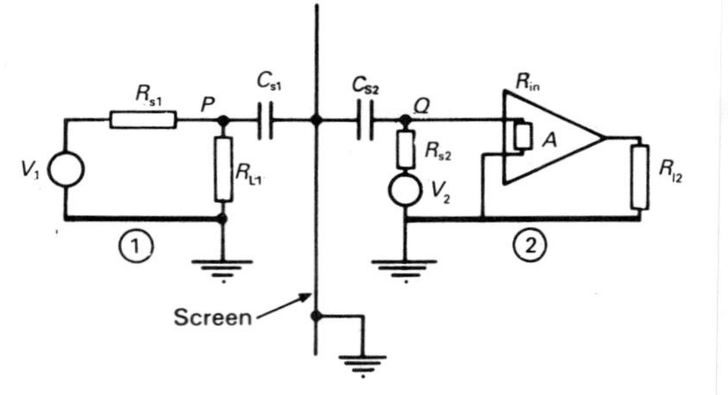

Fig. 2.5 The unwanted coupling between two circuits illustrated in Fig. 2.4 can be greatly reduced by putting an earthed screen between them. Any current passing through C_{s1} is conducted to earth instead of passing through C_{s2} to Q.

is earthed, otherwise P is connected to Q by C_{s1} in series with C_{s2} without a signal path to earth from their common point.

When very good screening against low-frequency electric fields is required then a closed conducting box must be used. Since all the sources of the field lie outside the box and it must be an equipotential surface it follows that, theoretically, the electric field inside it is zero. Such a box is sometimes known as a 'Faraday cage'. In practice, any enclosure will have one or more holes in it to allow wires to pass in and out and these may reduce the effectiveness of the screen. Similarly, the screening effectiveness is affected by the way in which the joints of the box are made. The **electric screening effectiveness** of an enclosure is defined as the ratio of the magnitude of the electric field with the screen to that at the same point when the screen is removed. It is usually expressed in decibels. At high frequencies other factors come into play and these are discussed in Chapter 9.

The use of differential amplifiers in this way is discused in Horrocks, D.H. *Feedback Circuits and Op Amps*, Van Nostrand Reinhold (1983).

The effects of capacitive coupling can also be reduced by using a differential amplifier if the source V_2 can be isolated from earth. Figure 2.6 shows how this works. The stray capacitances C_{s1} and C_{s2} are often approximately equal and so the spurious signals appearing at the normal and inverting inputs of the amplifier are also very nearly the same. If the amplifier has a high common-mode rejection ratio, then only their difference is amplified and added to the wanted signal. Adding an earthed screen reduces the stray capacitive coupling and therefore reduces the unwanted signal still further. The source for circuit 2 may be a transducer situated some distance from the amplifier. In that case the connecting cables must be screened as well as the amplifier, a point which is discussed further in Chapter 6.

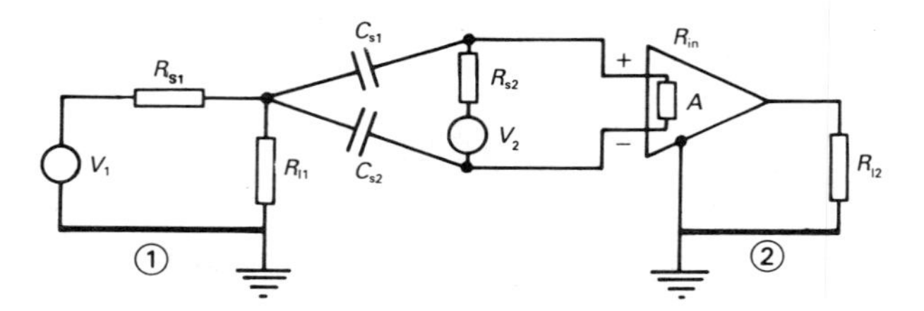

Fig. 2.6 When the input to a circuit can be isolated from earth it is possible to reduce capacitive coupling effects by using an amplifier with a differential input. The unwanted signals coupled through C_{s1} and C_{s2} are approximately equal but only their difference is added to the signal.

It is also possible for the capacitive coupling to be between the output and the input of an amplifier. When that happens there can be positive feedback causing the circuit to oscillate, so once again screening is needed.

Calculation of capacitance

In simple cases where the field problem can be solved by the direct application of Gauss' theorem it is straightforward to calculate the capacitance. The sequence of steps is:

1. Assume charges $\pm q$ on the conductors.
2. Apply Gauss' theorem, superposition or the method of images to find $\boldsymbol{E}$.
3. Integrate $\boldsymbol{E}$ along any convenient path between the conductors to find the potential difference between them.
4. Calculate the capacitance from $C = q/V$.

The potential difference between the conductors must be the same regardless of the path of integration, so we can choose the path which makes the integration as easy as possible.

Worked Example 2.3

Find the capacitance per unit length of the coaxial system shown in Fig. 2.3.

Solution Using the results of Worked Example 2.2,

$$C = 2\pi\left(\frac{1}{\epsilon_1}\ln(b/a) + \frac{1}{\epsilon_2}\ln(c/b)\right)^{-1}$$

Worked Example 2.4

Figure 2.7 shows the cross-section of two adjacent tracks on a printed circuit board. If the tracks run parallel to each other for 50 mm, estimate the capacitance between them. The relative permittivity of the material of the board is 6.0.

Solution The exact solution to this problem would be rather difficult, but an adequate estimate can be obtained by elementary methods. Consider first the approximation of the system by two parallel cylindrical wires as shown in the figure in the margin. We should not be too far out if we take the separation of the conductors to be the same in each case and the diameters of the wires to be equal to the widths of the strips on the printed circuit board. From Worked Example 1.4 the electric field between the wires in their plane is given by

Solutions to this and similar problems concerning parasitic capacitance can be found in Walker.*

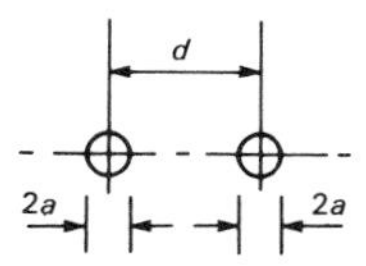

$$E = \frac{q}{2\pi\epsilon_0(x - d/2)} - \frac{q}{2\pi\epsilon_0(x + d/2)}$$

where charges $\pm q$ per unit length have been assumed to exist on the wires.

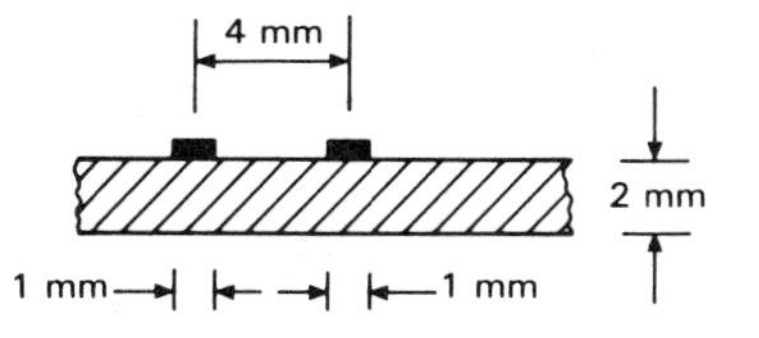

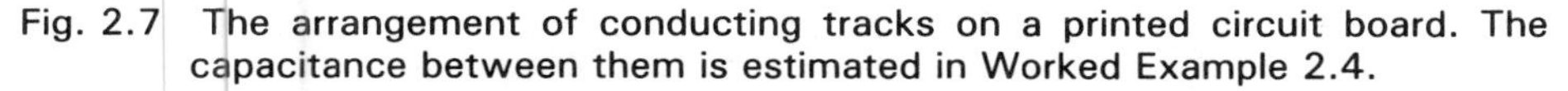

Fig. 2.7 The arrangement of conducting tracks on a printed circuit board. The capacitance between them is estimated in Worked Example 2.4.

Integrating the electric field from $x = -\frac{1}{2}d + a$ to $x = \frac{1}{2}d - a$ we obtain the potential difference between the wires as

$$V = \frac{q}{\pi\epsilon_0} \ln\left(\frac{d-a}{a}\right)$$

The capacitance per metre of the parallel wires in free space is then

$$C = q/V = \frac{\pi\epsilon_0}{\ln((d-a)/a)}$$

So, for the dimensions given, $C = 0.05\,\pi\epsilon_0/\ln(3.5/0.5) = 0.7$ pF. If the wires were wholly embedded in the dielectric, then the capacitance would be this figure multiplied by the relative permittivity, that is 4.2 pF. Now the capacitance between the conductors can be regarded as two capacitances in parallel, one for the part of the diagram above the board (C_1) and the other (C_2) for the part within and below it. Since capacitances in parallel add, C_1 must be just half the capacitance between the wires in free space, so that $C_1 = 0.35$ pF. The value of C_2 must lie somewhere between this figure and half of the capacitance between the wires when they are wholly embedded in the dielectric, thus $0.35\text{ pF} < C_2 < 2.1\text{ pF}$. Combining these estimates for C_1 and C_2 gives for the total capacitance

$$0.7\text{ pF} < C < 2.5\text{ pF}$$

Although this estimate is rather crude it is probably accurate enough for the purpose for which it was required. This calculation is typical of many that are made by engineers to get rough values of parameters.

When the shapes of the electrodes are more complicated than those in the examples above it is no longer possible to use the same method to calculate the capacitance. An alternative method is:

These steps can be carried out using numerical methods if analytical methods cannot be used.

1. Assume potentials 0 and V on the electrodes.
2. Solve Laplace's equation to give values of the potential everywhere.
3. Calculate $\mathbf{E}$ close to the surface of one of the electrodes using $\mathbf{E} = -\text{grad } V$.
4. Calculate the surface charge density distribution on the electrode using $\sigma = \epsilon E$.
5. Integrate the charge density distribution over the surface of the electrode to give the total charge q.
6. Calculate the capacitance from $C = q/V$.

Worked Example 2.5 Calculate the capacitance per unit length for the coaxial system of electrodes shown in the diagram in the margin.

4a

2a

2a

Solution Figure 2.8 shows the potentials at the mesh points close to the outer electrode taken from Fig. 1.15(b). The potential difference across AB is 23.9 V so the electric field strength is $(23.9/\frac{1}{4}a)\text{ V m}^{-1}$. The charge density at A is therefore $(95.6\,\epsilon_0/a)\text{ C m}^{-2}$. Assuming that this is the mean charge density between R and S, the total charge on a strip of the outer conductor lying between R and S and having unit length is $Q = (95.6\,\epsilon_0/a)(\frac{1}{4}a) = 23.9\,\epsilon_0\text{ C m}^{-1}$. The charges on each section of the outer conductor can be calculated in this way. The results of the calculations are shown in Fig. 2.8. Summing these charges gives the total charge

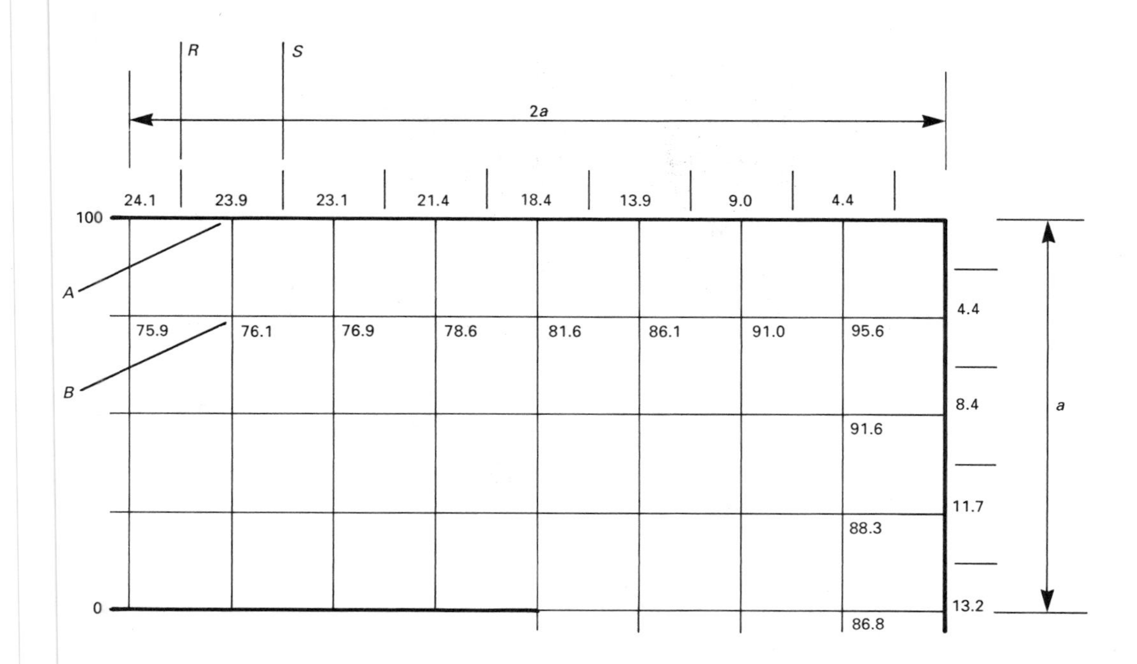

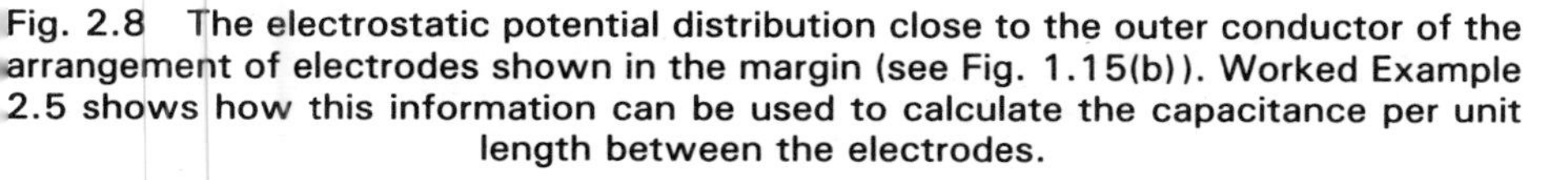

Fig. 2.8 The electrostatic potential distribution close to the outer conductor of the arrangement of electrodes shown in the margin (see Fig. 1.15(b)). Worked Example 2.5 shows how this information can be used to calculate the capacitance per unit length between the electrodes.

on the outer conductor: $629\,\epsilon_0\,\mathrm{C\,m^{-1}}$ for a potential difference of 100 V between the electrodes. The capacitance per unit length is therefore $6.3\,\epsilon_0\,\mathrm{F\,m^{-1}}$.

When this figure is compared with the exact value ($5.87\,\epsilon_0$) it is found to be in error by about 7%. This is accurate enough for many purposes but greater accuracy can be obtained, if required, by using a finer mesh.

The exact figure is given by Sykulski in a paper published in *IEE Proceedings*, Vol. 135, pp. 145–50 (1988).

To summarize: capacitance can be obtained by calculating either

1. the potential difference between the electrodes from assumed charges, or
2. the charges on the electrodes from an assumed potential difference between them.

By using both approaches in an approximate way it is possible to obtain upper and lower bounds to the capacitance. The formal application of this in the energy method is described later in the chapter. In the example which follows the two estimates are obtained using physical arguments.

Worked Example 2.6

Use approximations to the equipotential surfaces and field lines to obtain estimates of the capacitance per unit length of the electrodes described in Worked Example 2.5.

Solution From the finite difference solution to this problem shown in Fig. 1.15(b) we can plot the equipotentials, field lines and charge distributions as shown in Fig. 2.9.

If conducting sheets were placed along the equipotentials the field pattern would

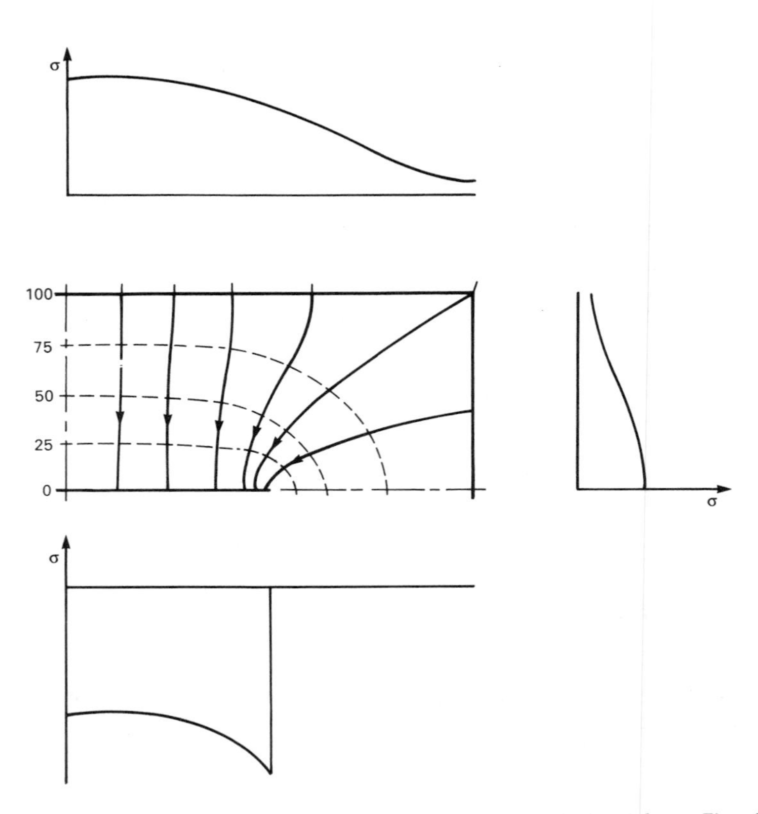

Fig. 2.9 The field map and charge density distributions derived from Fig. 1.15(b).

remain the same. It follows that the whole capacitance can be thought of as being made up of a set of capacitors in series, with each capacitor being bounded by a pair of adjacent equipotentials. This idea can be used to estimate the capacitance from a set of approximate equipotential surfaces. These surfaces must be chosen so that the electrodes form part of the set. Figure 2.10(a) shows one possible way of satisfying this requirement. Now consider the elementary capacitor which is formed by the surfaces which are x and $x + dx$ from the inner electrode. The perimeter of this capacitor is $(4a + 8x)$ so that

$$d\left(\frac{1}{C}\right) = \frac{dx}{\epsilon_0(4a + 8x)}$$

When this is integrated from $x = 0$ to $x = a$ the result is

$$1/C = (\ln 3)/8\epsilon_0$$

so that $C = 7.3\,\epsilon_0$. Comparison between Figures 2.9 and 2.10(a) shows that the potential gradient close to the outer electrode is greater in the latter case. Thus the total charge associated with the approximate equipotentials is greater than the true value. We therefore conclude that the actual capacitance must be less than the figure above.

In the second approach we imagine the charges and their associated field lines being rearranged as shown in Fig. 2.10(b). Because the charges are crowded more

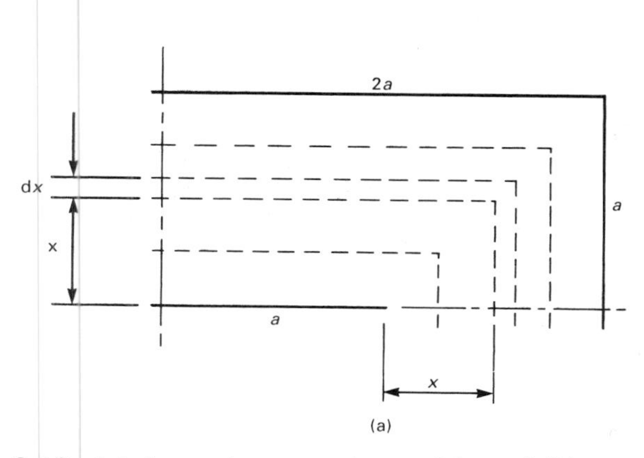

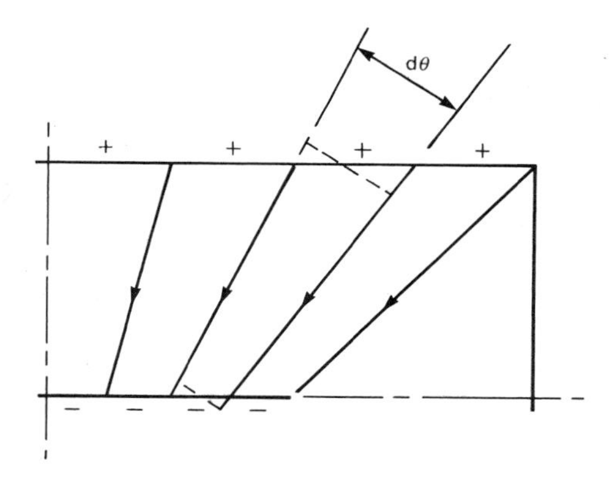

Fig. 2.10 (a) Approximate equipotentials and (b) approximate field lines for the problem discussed in Worked Example 2.6.

closely together than in their equilibrium state we expect the potential difference between the electrodes to be increased so that the estimate of the capacitance will be lower than the actual value. In this case we can imagine the whole capacitance as being made up of a number of wedge-shaped capacitors in parallel. Each elementary capacitor is bounded by a pair of field lines and can be thought of as a sector of a coaxial capacitor. The boundaries at the ends of the element are small arcs which do not coincide exactly with the electrode surfaces but, provided that $\mathrm{d}\theta$ is small, the error is not very large. The capacitance of the element is readily deduced from the equations in Worked Example 2.1 to be

$$\mathrm{d}C = \epsilon_0 \mathrm{d}\theta / \ln\,(b/a)$$

where a and b are the inner and outer radii of the element, respectively. Since the charge distributions on the electrodes have been assumed to be uniform it follows from the geometry of the problem that $b = 2a$ for every element. The equation above is then readily integrated from $\theta = 0$ to $\theta = \pi/4$ to find the capacitance of one quadrant and the whole capacitance is therefore

$$C = \pi\epsilon_0 / \ln 2 = 4.5\,\epsilon_0$$

If we take the arithmetic mean of the two values of capacitance as the best estimate, we conclude that $C = (5.9 \pm 1.4)\,\epsilon_0\,\mathrm{F\,m^{-1}}$. This figure is about 0.5% greater than the exact value ($5.87\,\epsilon_0$) and is actually much closer to it than the result of the finite difference calculation. It should not be expected that the mean of the upper and lower bounds calculated in this way will always be as accurate as this. Nevertheless, the method usually gives accuracy which is remarkable considering the crudity of the assumptions made.

Energy storage in the electric field

When charge is transferred from one plate of a capacitor to the other work is done against the electric field. This work is stored as potential energy in the capacitor. It is easy to show that the stored energy is given by

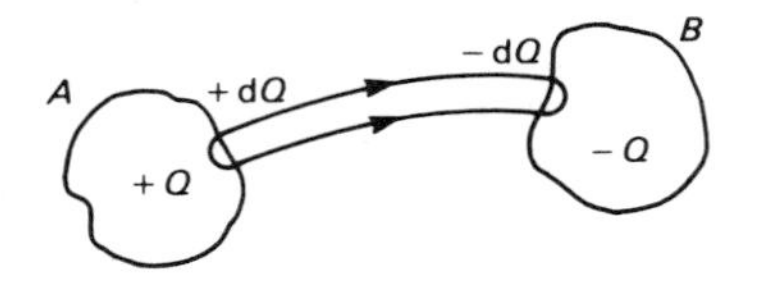

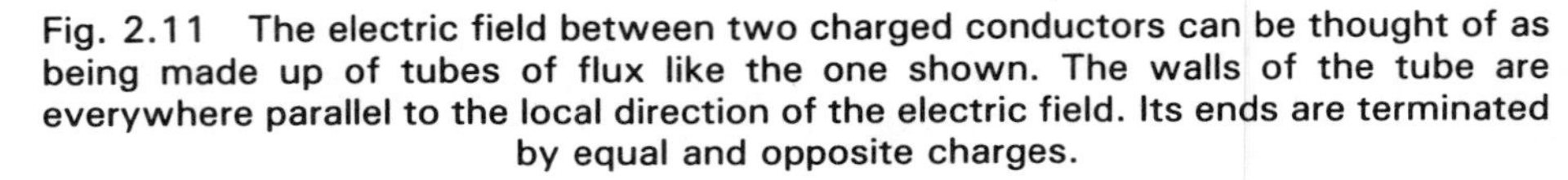
Fig. 2.11 The electric field between two charged conductors can be thought of as being made up of tubes of flux like the one shown. The walls of the tube are everywhere parallel to the local direction of the electric field. Its ends are terminated by equal and opposite charges.

The proof of this equation is given by Compton.*

$$W = \tfrac{1}{2}Q^2/C = \tfrac{1}{2}CV^2 = \tfrac{1}{2}QV \tag{2.9}$$

It is important to remember that Equation (2.9) applies to any pair of electrodes with a potential difference between them, not just to capacitors as lumped components.

It is sometimes useful to think of the energy stored in a capacitor as being distributed throughout the electric field associated with it. Figure 2.11 shows a capacitor with electrodes of arbitrary shape. We define a **flux tube** by considering a small element of the surface of the conductor A and following the lines of $\boldsymbol{E}$ which start from its boundary through space until they terminate upon B. Since the flux lines can never meet or cross each other it follows that the whole electric field of the capacitor can be divided up into flux tubes. In addition we observe that, if the charge on the element of A from which the tube starts is $+\mathrm{d}Q$, then that on the element of B on which it ends must be $-\mathrm{d}Q$ by a straightforward application of Gauss' theorem to a Gaussian surface enclosing the tube. The figure in the margin shows a small part of the flux tube on a larger scale. A short length of the tube is chosen having length $\mathrm{d}l$ as shown, and the cross-sectional area of the tube at this point is $\mathrm{d}A$. If the local electric field strength is E, then the potential difference between the ends of this element of volume is $\mathrm{d}V = E\,\mathrm{d}l$, where the dot product can be omitted because $\boldsymbol{E}$ and $\mathrm{d}\boldsymbol{l}$ are parallel to each other by definition. Likewise, the relationship between the electric flux in the tube and the charges at its ends means that $\mathrm{d}Q = D\,\mathrm{d}A$. The energy stored in the element of volume is then

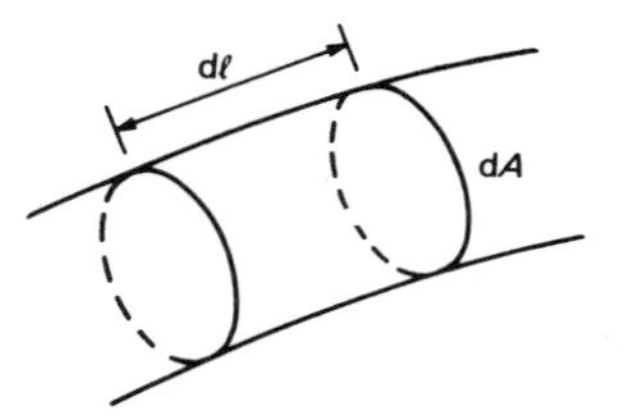

$$\mathrm{d}W = \tfrac{1}{2}\mathrm{d}Q\,\mathrm{d}V = \tfrac{1}{2}DE\,\mathrm{d}l\,\mathrm{d}A = \tfrac{1}{2}DE\,\mathrm{d}v$$

using Equation (2.9) with the volume of the element represented by $\mathrm{d}v$. In the limit as $\mathrm{d}v$ shrinks to zero the energy density in the electric field is $\tfrac{1}{2}DE$. A more rigorous derivation of this result which allowed for the possibility that the vectors $\boldsymbol{D}$ and $\boldsymbol{E}$ are not parallel to each other would show that the energy density in the field is given by

$$w = \tfrac{1}{2}\boldsymbol{D}\cdot\boldsymbol{E} \tag{2.10}$$

and the total energy stored in the field of the electrodes is

$$W = \tfrac{1}{2}\iiint \boldsymbol{D}\cdot\boldsymbol{E}\,\mathrm{d}v \tag{2.11}$$

where the integral is taken over all that part of space in which the electric field is not zero. In the worked example which follows it is shown that the same value for the stored energy is given by Equations (2.9) and (2.11).

Worked Example 2.7

Calculate the stored energy per unit length in the coaxial cable shown in Fig. 2.2 when the voltage between the electrodes is V_0, using Equations (2.9) and (2.11), and compare the results.

Solution The capacitance per unit length of the cable is

$$C = 2\pi\epsilon / \ln(b/a)$$

from the result of Worked Example 2.3. Then, using Equation (2.9), the stored energy per unit length is

$$W = \pi\epsilon V_0^2 / \ln(b/a)$$

Alternatively, the electric field at radius r is given by

$$E = V_0 / [r \ln(b/a)]$$

from Worked Example 2.1. The stored energy in a thin cylinder of thickness $\mathrm{d}r$ at radius r is then

$$\mathrm{d}W = \tfrac{1}{2} DE \, 2\pi\epsilon \, \mathrm{d}r = \frac{\pi\epsilon V_0^2}{(\ln(b/a))^2} \frac{dr}{r}$$

Integrating over r from $r = a$ to $r = b$ gives

$$W = \pi\epsilon V_0^2 / \ln(b/a)$$

as before.

Calculation of capacitance by energy methods

This is a rather simplified discussion of the method. A much more rigorous treatment is given by Hammond.*

In many cases the solution required to a field problem is not detailed field information but a single number, the capacitance. The use of ideas of stored energy provides a particularly simple way of obtaining estimates of capacitance without finding the exact solution to the field problem. The starting point is the fact that when a physical system is in stable equilibrium its stored energy is a minimum. From this it follows that any perturbation of the system must result in an increase in the stored energy. Consider now two possible perturbations of the system of charges and electric fields in a capacitor. In the first case we retain the electrodes as equipotential surfaces but alter the charges on them in such a way that the equipotentials in the inter-electrode space assume particularly simple shapes. From these perturbed equipotentials we calculate the energy density in the field and, hence, the stored energy W' of the system. This energy must be greater than the stored energy W of the unperturbed system, so that

$$W' \geqslant \tfrac{1}{2} C V^2 \qquad \text{or} \qquad C \leqslant 2W'/V^2 \tag{2.12}$$

This provides an upper bound to the capacitance.

In the second case we hold the charges on the electrodes constant but redistribute them in such a way that the flux lines have a particularly simple form. When this is done the electrodes are no longer equipotential surfaces. Once again the stored energy, say W'', is calculated, and this time

$$W'' \geqslant Q^2/2C \qquad \text{or} \qquad C \geqslant Q^2/2W'' \tag{2.13}$$

Notice how the upper and lower bounds arise naturally from the two alternative strategies for calculating capacitance which were described earlier.

giving a lower bound for the capacitance.

The upper and lower bounds for the capacitance obtained in Worked Example 2.6 were based upon approximate equipotential surfaces and flux lines. In the example a physical argument was used to justify the assertion that the values of capacitance obtained were indeed upper and lower bounds. We have now shown that the assertion can be justified rigorously.

See Sykulski, J.K., Computer package for calculating electric and magnetic fields exploiting dual energy bounds, *IEE Proceedings*, Vol. 135, pp. 145–50 (1988).

In theory, it is possible to use the method of Worked Example 2.6 to obtain better estimates of the capacitance by choosing more complicated approximations to the field pattern. In practice, the effort involved in doing this by hand is not justified by the improvement in accuracy. The computer method of tubes and slices makes it possible to apply the technique to problems with more complicated boundary conditions. This method and its application to the problem of Worked Example 2.6 are described by Sykulski.

Finite element method

An excellent introduction to this subject is given by Silvester and Ferrari.*

Many computer packages for the solution of field problems are based upon the use of finite elements. This technique is not suitable for hand calculation and the algorithm is a little complicated so only the principles of the method as applied to two-dimensional problems are described here.

In the finite element method the region between the electrodes is divided into a large number of triangular elements. Figure 2.12(a) shows one such element in the x–y plane. It can be shown that the condition that Laplace's equation should be obeyed in the element is equivalent to the requirement that the energy stored within it should be a minimum. If the potential within the element is approximated by

$$V(x, y) = a + bx + cy \tag{2.14}$$

then the electric field components are

$$E_x = -b \qquad \text{and} \qquad E_y = -c \tag{2.15}$$

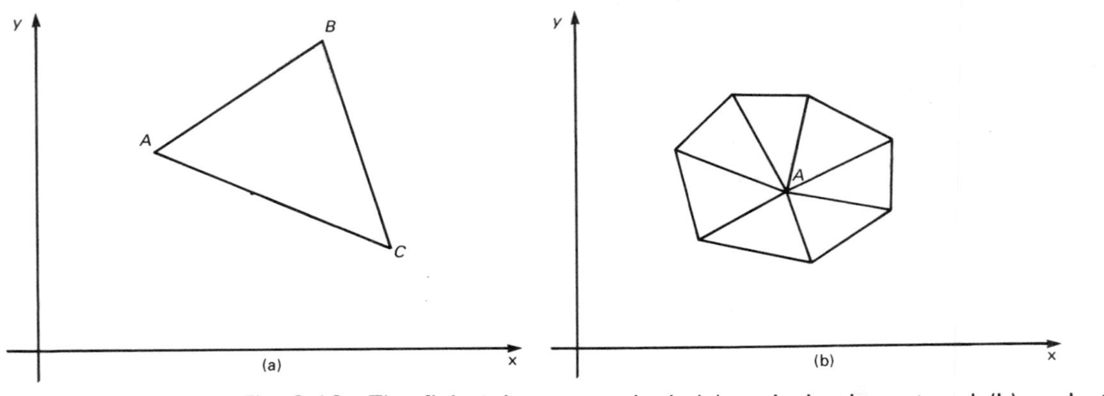

Fig. 2.12 The finite element method: (a) a single element and (b) a cluster of elements.

and the energy density is

$$w = \tfrac{1}{2}\epsilon_0(b^2 + c^2) \tag{2.16}$$

Now the coefficients b and c can be expressed in terms of the positions of the vertices of the triangle and of the potentials at them. Thus the stored energy in the element can be expressed in terms of the same quantities.

Now consider the cluster of triangular elements with a common vertex at A, as shown in Fig. 2.12(b). The energy stored in the cluster of elements, W, is the sum of the energies stored in each. We now select the value of the potential at A which minimises the stored energy by requiring that

$$\partial W/\partial V_A = 0 \tag{2.17}$$

The result is an equation giving the potential at A in terms of the potentials at the other vertices and their positions. It is interesting that, if triangular elements are formed from the square elements in Fig. 1.14 by adding diagonals from the lower left to the upper right corner of each square, then the application of the finite element method yields Equation (1.30).

The finite element method is important chiefly because of the freedom with which the sizes and shapes of the elements may be chosen. It is, therefore, possible to select a set of elements which fits the boundaries of the problem with good accuracy and which provides a concentration of small elements in regions where the potential is changing rapidly. In two-dimensional problems the spatial variation of the potential can be represented as a smooth surface. The finite element method approximates this surface by one with many small flat facets rather like the surface of a cut gem.

Boundary element method

The finite difference and finite element methods both depend upon the division of the space between the electrodes into a large number of small elements. The potential values at the nodes of the mesh are manipulated according to some algorithm until the solution has converged. The result of the process, in each case, is detailed information about the field at every point within the solution space. In large three-dimensional problems the number of variables can be very large indeed and the computation time correspondingly long.

Very often the result required from a field calculation is not the detailed field information but only a single number such as the capacitance. For problems of this kind an alternative method, the boundary element method, is useful because it only requires mesh points and potentials to be specified on the boundaries. This has the benefit that the number of variables in the solution is less than in the corresponding finite difference or finite element solution by two or three orders of magnitude. It also makes it easier to model problems with complicated boundaries. The fields at internal points can be calculated from the solution at the boundary if required.

The application of boundary element methods to the solution of Laplace's equation is described by Brebbia, C.A. (1978) *The boundary element method for engineers*. Pentech Press, London*.

Summary

Insulating materials are widely used in electronic engineering both to provide electrical isolation and to enhance the performance of capacitors. The dielectric qualities of most insulating materials can be represented by a constant, the permittivity. It is convenient to introduce a new field vector, the electric flux density $\boldsymbol{D}$, whose value is independent of the presence of dielectric materials. The use of this vector makes the solution of problems involving dielectric materials easier. It is important to use the correct boundary conditions at the interfaces between different dielectric materials. Capacitance occurs in capacitors and in stray capacitance between conductors. It can be calculated either by a direct solution of the field problem or by considering the stored energy in an approximation to the field associated with the capacitor. The latter method gives quite good accuracy with relative ease. Stray capacitances lead to unwanted coupling between electronic circuits. Techniques for reducing the coupling can be understood using the theory of electrostatics.

Problems

You should find that the maximum working voltage is considerably greater when the air gap is eliminated. This is why high-voltage equipment is immersed in oil or potted in epoxy resin.

2.1 A capacitor consists of two parallel conducting plates whose area is large enough compared with their spacing for edge effects to be negligible. The sheets are 0.1 mm apart and the space between them is partially filled with polythene sheet of thickness 0.09 mm, relative permittivity 2.25 and breakdown strength $30\,\mathrm{MV\,m^{-1}}$. Assuming that the breakdown strength of air is $3\,\mathrm{MV\,m^{-1}}$, calculate the maximum voltage which can be applied to the capacitor. What difference does it make if the polythene sheet completely fills the space between the plates?

2.2 Show that the capacitance per unit length between the parallel wires shown in Fig. 1.6 is given by

$$C = \pi\epsilon_0 / \ln\left((d/a) - 1\right) \qquad \text{if} \quad d \gg a$$

2.3 The parallel wires of Problem 2.2 are placed so that each is $d/2$ from a large flat sheet of metal. How does this affect the capacitance? Calculate the capacitance per unit length between the wires in the presence of the metal sheet if $a = 1$ mm and $d = 20$ mm.

2.4 A coaxial cable is to be made with two dielectric layers as shown in Fig. 2.3. The inner layer is made from a high-quality but expensive material which has dielectric strength E_1 and the outer layer from a cheaper material of dielectric strength E_2. Find an expression for the outer radius, b, of the inner layer if the cable is to be made as cheap as possible without reducing its maximum working voltage.

2.5 A variable tuning capacitor for a radio set consists of a set of fixed plates, A, and a set of moving plates, B, as shown in Fig. 2.13. The frequency to which the radio is tuned varies inversely as the square root of the capacitance. Assuming that the effects of fringing fields can be neglected, find the shape which the moving plates must have if the frequency is to be proportional to the angle θ in the range 20–160° and 500–1500 kHz.

2.6 Calculate the capacitance per unit length of the square coaxial arrangement of electrodes shown in the figure in the margin:

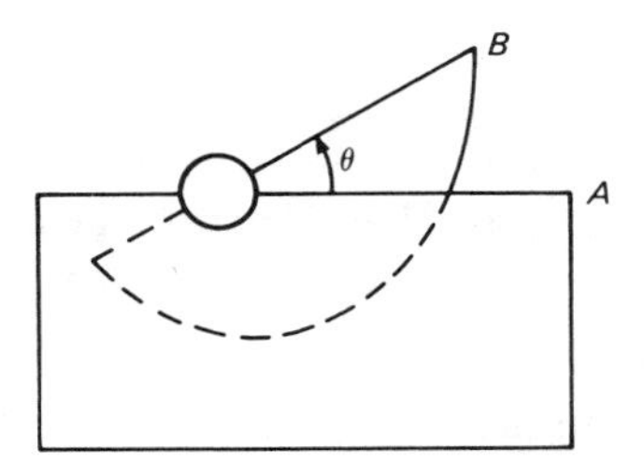

Fig. 2.13 A schematic diagram of the type of variable capacitor used for tuning radio sets. A set of moving plates B rotates within a parallel set of fixed plates A.

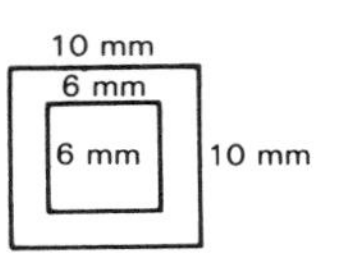

(a) by making use of the results of the finite difference solution obtained in Problem 1.7;
(b) by energy methods.

2.7 Figure 2.14 shows the distribution of electric charges in the depletion region of a p–n junction in equilibrium. Assuming that the silicon has a relative permittivity ϵ, solve Poisson's equation and show that the potential difference between the two sides of the junction is given by:

$$V = \frac{q}{2\epsilon}\,(N_D d_n^2 + N_A d_p^2)$$

where q is the magnitude of each charge.

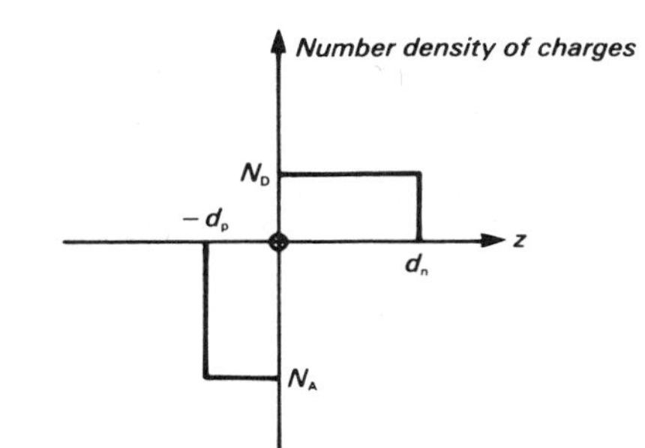

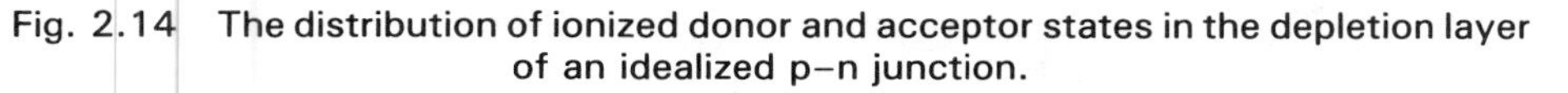

Fig. 2.14 The distribution of ionized donor and acceptor states in the depletion layer of an idealized p–n junction.

2.8 The formula obtained in Problem 2.7 is still valid when a reverse-bias voltage is applied to the junction with V replaced by the magnitude of the reverse bias V_A. If a small a.c. signal dV_A is superimposed on the bias voltage, the effective capacitance is given by

$$C_j = \frac{dQ}{dV_A}$$

where Q is the total charge on either side of the junction. The net charges on the two sides of the junction are always equal to each other. Find an expression for the capacitance C_j in terms of the voltage V_A and the constants of the material.

A full discussion of the capacitance of a reverse biased p–n junction is given by Bar-Lev.* This effect is put to use in varactor diodes to provide a capacitance whose value can be controlled by an applied voltage.

2.9 A MOS transistor is essentially a parallel-plate capacitor comprising a silicon substrate, a silicon dioxide insulating layer, and an aluminium gate electrode. If the silicon dioxide has relative permittivity 3.85 and dielectric strength 6.0×10^8 V m^{-1}, and the insulating layer is 0.1 μm thick, estimate the maximum voltage which can be applied to the gate electrode.

2.10 Using the results of Problem 2.9 calculate the maximum charge per unit area which can be induced in the semiconductor material. If there are 2.0×10^{18} atoms per square metre in the first layer of the silicon crystal, what proportion can be ionized by applying a voltage to the gate which is one-sixth of the breakdown voltage?

Steady electric currents 3

Objectives

- ☐ To introduce the ideas of current flow in electrical conductors and of electrical resistance.
- ☐ To show that Ohm's law is a special case of a more general relationship between current density and electric field.
- ☐ To derive expressions for the power dissipated per unit volume within a conductor.
- ☐ To show how the distribution of current within a conductor can be calculated.
- ☐ To introduce the continuity equation and to show how Kirchhoff's current law is a special case of it.
- ☐ To introduce the concept of the electromotive force of a source of electric power.
- ☐ To modify the equation for the line integral of the electric field around a closed path derived in Chapter 1 to allow for the presence of sources of electric power, and to show that the resulting equation is a generalization of Kirchhoff's voltage law.
- ☐ To show how the resistance of a conductor can be calculated directly, by the use of Laplace's equation, and by energy methods.

Conduction of electricity

In the preceding chapters we have studied problems involving charges which are at rest or moving in a vacuum. This is convenient for the formal exposition of electromagnetic theory, but most people meet electrical phenomena for the first time in the form of electric currents flowing in wires. A current can flow in any medium in which there are charges which are free to move. These **conduction charges** may be electrons, positively charged 'holes', or positive or negative ions, according to the material. They are in continuous random motion, with a distribution of velocities which depends upon the temperature of the material. They are also constantly colliding with each other and with the atomic structure of the material. When a conducting material is placed in an electric field the conduction charges are accelerated in the direction of the field. The velocity acquired is small compared with the average value of the random velocity at ordinary temperatures, and is superimposed on it. The ordered part of the motion would increase without limit were it not for the collisions which convert it into random motion. Overall, the effect of the field is to add a small average **drift velocity** to the random velocity. The magnitude of the drift velocity is related to the strength of the field by the equation

For a discussion of the motions of electrons and holes in semiconductor materials, see Bar-Lev.*

$$v_d = \mu E \tag{3.1}$$

where μ is known as the **mobility** of the charge carriers. The mobility depends on

1. the type of charge carrier;
2. the material in which it moves;
3. the temperature of the material.

It can also depend on

4. the strength of the electric field;
5. the orientation of the crystal axes to the field.

One very important example of a material in which the mobility of charge carriers depends upon the strength of the electric field is silicon. See Bar-Lev* for a discussion of velocity saturation and its causes.

For many materials μ is a scalar quantity which varies only with temperature. In this book we shall confine our attention to this case.

The rate of flow of electric charge across unit area of a plane normal to the direction of v_d is the **current density** given by

$$J = n\,q\,v_d \tag{3.2}$$

where n is the density of charge carriers and q the charge on each. Combining Equations (3.1) and (3.2) gives

$$J = n\,q\,\mu E = \sigma E \tag{3.3}$$

where σ is the **conductivity** of the material, measured in siemens per metre. J is measured in coulombs per square metre per second or in amperes per square metre, defining the **ampere** as a current of one coulomb per second. It is sometimes convenient to use the reciprocal of conductivity which is known as **resistivity** (ρ). Resistivity is measured in V m A^{-1} or Ω m.

The symbols ρ and σ are also used for charge densities, but both usages do not normally occur in the same problem so there is not much risk of confusion.

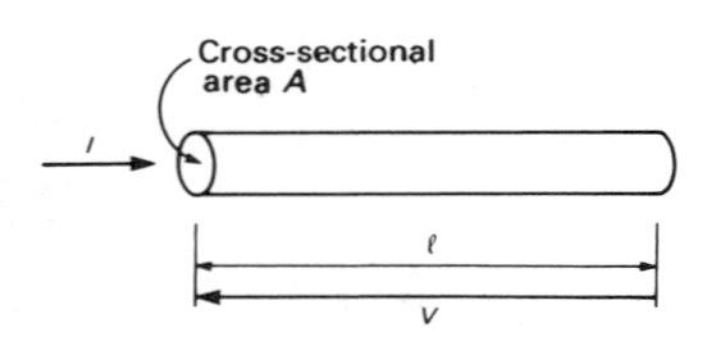

To link these ideas with those of elementary circuit theory consider the flow of electric current in a section of straight wire as shown in the figure in the margin. If the wire has cross-sectional area A and carries a current I uniformly distributed across it, then the current density is

$$J = I/A \tag{3.4}$$

The relationship between the electric field in the wire, the potential difference between its ends, and its length is

$$E = V/l \tag{3.5}$$

But $E = \rho J$ from Equation (3.3) and therefore

$$V = (\rho l/A)\,I = RI \tag{3.6}$$

which is the familiar form of Ohm's law. Equation (3.3) is thus shown to be a general form of Ohm's law applicable to problems in which the current density is not uniform.

Ohm's law is valid only if the mobility is independent of the electric field strength. Fortunately this is so in most materials except at very high field strengths.

Ohmic heating

When conduction charges are accelerated by an electric field, they continually gain energy from it. This ordered kinetic energy is transferred to the bulk of the material by collisions, so increasing the random thermal motions of the atoms.

The conversion of electrical energy into heat is familiar from its everyday use in electric heaters and light bulbs. For a steady current I sustained by an applied voltage V the power input is

In a lamp bulb the resistance of the filament increases with temperature. This tends to stabilize the temperature of the filament.

$$P = VI = I^2R = V^2/R \tag{3.7}$$

an equation which is well known from elementary circuit theory. By using the results of the previous section we can find corresponding formulae for the details of the energy dissipation within a conducting material. Consider the element of volume shown in the figure in the margin. Provided that the properties of the material are the same in all directions, we can set the x-axis parallel to the direction of the current without loss of generality. The potential difference across the element is

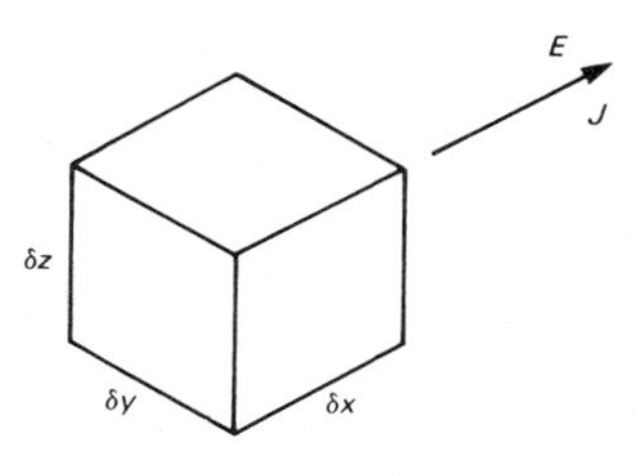

$$V = E\,\delta x \tag{3.8}$$

and the current flowing though it is

$$I = J\,\delta y\,\delta z \tag{3.9}$$

Thus the power dissipated in the element is

$$\delta P = EJ\,\delta x\,\delta y\,\delta z = EJ\,\delta v = \sigma E^2\,\delta v = \rho J^2\,\delta v \tag{3.10}$$

where δv is the volume of the element. A more rigorous derivation would show that

$$\delta P = \boldsymbol{E}\cdot\boldsymbol{J}\,\delta v \tag{3.11}$$

This formula is known as Joule's law. For a more detailed discussion consult Reitz and Milford.*

This expression is valid even when the vectors $\boldsymbol{E}$ and $\boldsymbol{J}$ are not parallel to each other. Equation (3.10) shows that the power density is greatest in regions of high current density. This phenomenon is of great practical importance. In fuses the provision of a section of thin wire in a circuit ensures that that section is heated to melting point before any other part of the circuit is damaged. On the other hand, wiring joints which are carelessly designed or made can overheat and fail.

The distribution of current density in conductors

We would like to know how to calculate the variation of current density with position. To do this we consider the net current flow out of a closed surface S enclosing a volume V as shown in the figure in the margin. If the current density in the region of S is $\boldsymbol{J}$ (not necessarily a constant) then the net current flow out of S is given by the integral of $\boldsymbol{J}$ over S, that is

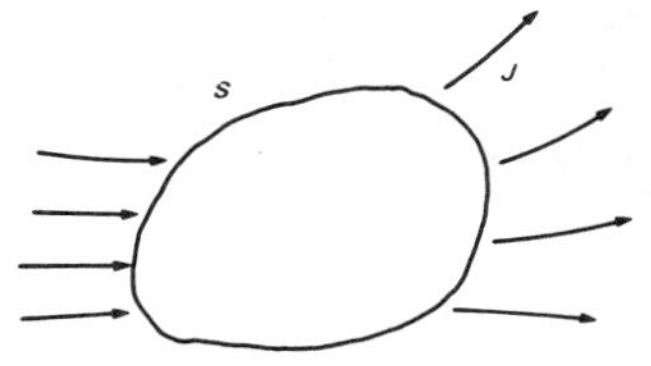

$$\text{current out of } S = \oint\!\!\oint_S \boldsymbol{J}\cdot\mathbf{d}\boldsymbol{A} \tag{3.12}$$

This is a mathematical statement of the principle of conservation of charge.

Now if there is a net current flow out of a closed surface, then the total charge enclosed by that surface must be changing with time because charges cannot be created or destroyed. Thus

$$\text{current out of } S = -\frac{\partial}{\partial t}\,(\text{charge enclosed by } S)$$

which may be written in mathematical notation as

$$\oint\!\!\oint_S \boldsymbol{J}\cdot\mathbf{d}\boldsymbol{A} = -\frac{\partial}{\partial t}\iiint_v \rho\,\mathrm{d}v \tag{3.13}$$

Note that ρ is charge density, not resistivity, in this equation.

Using the same method as we used in Chapter 1 to get from Equation (1.5) to Equation (1.11), it is easy to show that the differential form of Equation (3.13) is

$$\nabla \cdot \boldsymbol{J} = -\frac{\partial \rho}{\partial t} \tag{3.14}$$

In the theory of semiconductor materials it is necessary to write separate continuity equations for the electrons and the holes. These equations include an extra term to account for changes in charge density through generation and recombination. See Bar-Lev* for details.

This equation, in either its integral or its differential form, is known as the **continuity equation**. It is valid for the flow of charges in a vacuum as well as in conducting materials.

When the current flow is in a steady state the right-hand sides of Equations (3.13) and (3.14) become zero. If, furthermore, the conductivity of the material is constant, then

$$\nabla \cdot \boldsymbol{J} = \nabla \cdot \sigma \boldsymbol{E} = \sigma \nabla \cdot \boldsymbol{E} = 0 \tag{3.15}$$

but

$$\boldsymbol{E} = -\nabla V \qquad \text{(Equation (1.23))}$$

so that

$$\nabla^2 V = 0$$

Thus when a steady electric current flows in a material of constant conductivity the potential distribution obeys Laplace's equation.

In the special case of currents flowing in wires, as shown in Fig. 3.1, the integral on the left-hand side of Equation (3.13) becomes a summation over the currents in the wires

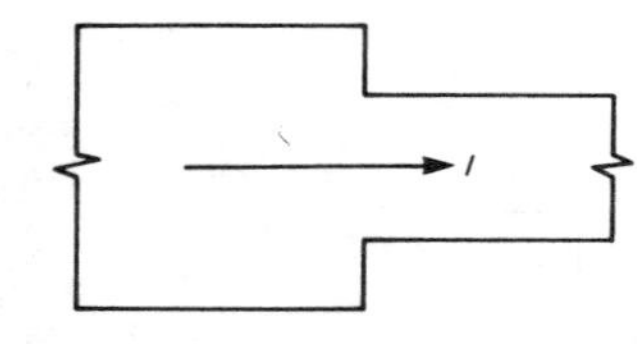

$$I_1 + I_2 + I_3 + \ldots + I_n = 0 \tag{3.16}$$

which is Kirchhoff's current law familiar from electrical circuit theory. The continuity equation is therefore a generalization of Kirchhoff's current law.

This problem can also be solved by conformal mapping. Tables of solutions obtained by this technique are given by Binns and Lawrenson.*

In general, provided that the conductivity is constant, the current density distribution can be calculated from the solution of Laplace's equation. The figure in the margin shows a typical problem: a strip with a step change in its width. In this case Laplace's equation would have to be solved by numerical methods. The potential is specified by assuming that at some distance from the step the equipotentials are straight lines perpendicular to the length of the strip. Along the sides there is a boundary between a conductor and an insulator, as shown in the figure in the margin. The potential cannot change abruptly as the boundary is crossed, so the tangential component of the electric field must be the same on both sides

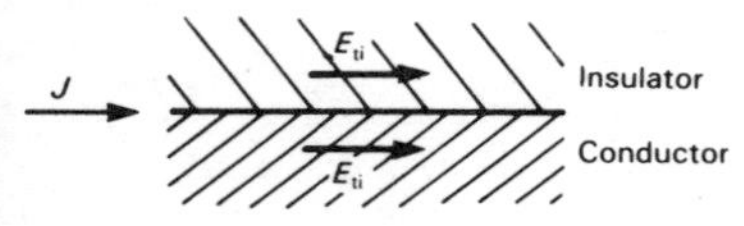

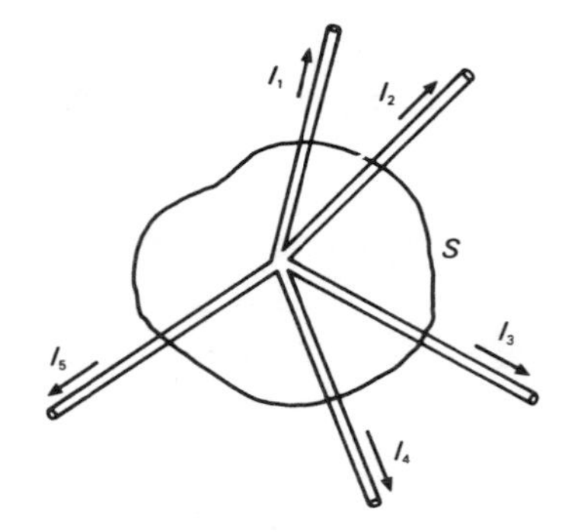

Fig. 3.1 Applying the continuity equation to the special case of the steady currents flowing in wires.

of it. If the current is reduced to zero, then $E_{tc} = 0$, so that $E_{ti} = 0$ also and the electric field in the insulator is normal to the conducting boundary, as it must be in electrostatic problems.

The connection between the electric field and the current density patterns was used in analogue methods for the solution of electrostatic problems. These methods, which can still be useful on occasion, involve modelling the electrostatic problem by current flow in a medium of low conductivity such as tap water or resistive paper. The equipotential surfaces are plotted by measuring the potential at points within the conducting medium using bridge methods or a high-impedance voltmeter.

A detailed discussion of electrolytic tank methods is given by Einstein, P.A., Factors limiting the accuracy of electrolytic plotting tanks, *Brit. J. App. Phys.* **2**, 49 (1951).

Electric fields in the presence of currents

When electric currents flow through a system of conductors there is an electric field in the space around the conductors besides the field within the conductors driving the current. As an example, consider the field around a coaxial line carrying equal and opposite currents in the core and the sheath, as shown in Fig. 3.2. Within the conductors the field lines, like the current flow lines, lie parallel to the axis. Outside them the field is modified from the purely radial electrostatic solution. The finite resistance of the conductors requires that the potential at A should be higher than that at B and that at C should, likewise, be greater than that at D. There must, therefore, be axial components of the electric field in the space between the conductors as well as in the conductors themselves. There is also a radial component of the electric field in the space between the conductors because V_A is greater than V_D and V_B is greater than V_C. The magnitude of the radial field component varies along the length of the line because $(V_A - V_D)$ is greater than $(V_B - V_C)$. The field pattern resulting from the combination of the axial and radial fields is shown in Fig. 3.2. The difference from the electrostatic case is that the field lines and equipotentials are no longer straight but curved, as shown.

As another example, consider the change which occurs in the electric field around a circuit when the current in it is interrupted by a switch. Figure 3.3(a) shows a simple circuit containing only a battery and a resistor. The field pattern around the circuit is roughly as shown, provided that the resistance of the connecting wire is much less than that of the resistor. When the circuit is broken there is a dramatic change in the field. There is now no component of the field parallel to the axis of the resistor because there is no current flowing in it. The field is concentrated around the open switch, as shown in Fig. 3.3(b).

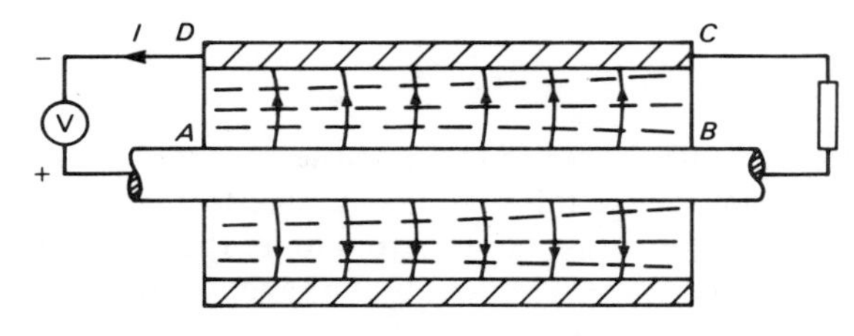

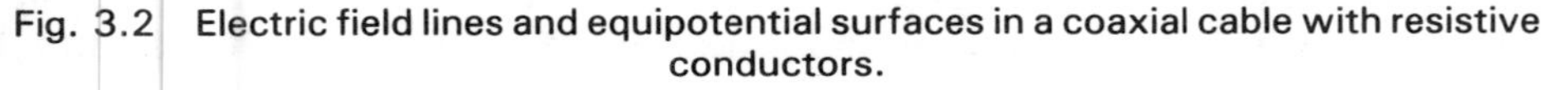
Fig. 3.2 Electric field lines and equipotential surfaces in a coaxial cable with resistive conductors.

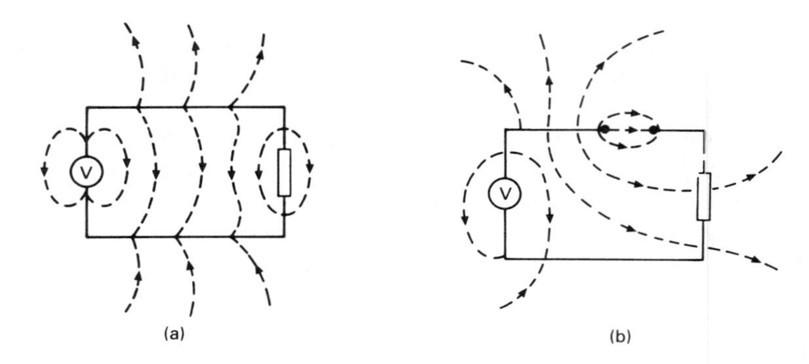

Fig. 3.3 The electric field lines around a simple circuit: (a) when current is flowing; (b) when the circuit is broken.

Electromotive force

For a detailed discussion of electromagnetic generators, see Fitzgerald, A.E., Kingsley, C. and Umans, S.D., *Electric Machinery* (4th edn), McGraw-Hill (1983). For electric cells, see Barak, M. (ed.), *Electrochemical Power Sources: primary and secondary batteries*, Peter Peregrinus (1980).

The voltage and current generators of a.c. circuit theory are actually idealizations of the effects of variable impedances on currents from d.c. sources.

In this chapter we have so far assumed that sources of electric current exist, but have paid no further attention to them. These sources have been represented as idealized voltage generators. In practice, they are most likely to be electrochemical cells or electromagnetic generators. The details of the inner workings of these sources do not fall within the scope of this book, but it is important to consider the effects of sources in the theory of electromagnetism. We shall regard a source as a device capable of maintaining a potential difference and a steady flow of electric current between its terminals. Such a device is a source of electric power because it does work in driving the electric current around the circuit against the circuit resistance. As the charges move around the circuit from one terminal of the source to the other, they lose potential energy continuously as it is converted into heat.

In Chapter 1 we saw that in an electrostatic field the line integral of the electric field around a closed path is zero

$$\oint \boldsymbol{E} \cdot \mathbf{d}\boldsymbol{l} = 0 \qquad \text{(Equation (1.16))}$$

This equation remains true near a current-carrying circuit provided that the path of integration does not pass through the generator. But if the line integral of the electric field is evaluated around the circuit from one terminal of the source to the other, the result is not zero. This is because a net amount of work has to be done to move the charges along this path against the resistance of the circuit. The potential energy given up by a charge in this way is restored when it moves through the source. This can be expressed mathematically by writing

$$\int_C \boldsymbol{E} \cdot \mathbf{d}\boldsymbol{l} = \mathcal{E} \tag{3.17}$$

Thévenin's theorem and its applications are discussed by Carter, G.W. and Richardson, A., *Techniques of Circuit Analysis*, Cambridge University Press (1972).

where $\mathcal{E}$ is the **electromotive force** (e.m.f.) of the source and the integral is taken around the circuit from one terminal of the source to the other.

When the source is imperfect it is usual to equate the electromotive force with the **Thévenin voltage**, that is the open-circuit voltage, of the source and to include the source impedance in the external circuit. For the special case of a circuit consisting of lumped components and generators joined by lossless conductors, the integral in Equation (3.17) becomes the sum of the potential differences across the components. The electromotive force is, likewise, the sum of the e.m.f.s of all the generators in the circuit. The result is Kirchhoff's voltage law:

$$R_1I_1 + R_2I_2 + R_3I_3 + \ldots + R_nI_n = \mathcal{E}_1 + \mathcal{E}_2 + \ldots + \mathcal{E}_m \qquad (3.18)$$

Calculation of resistance

From Ohm's law in the form

$$V = RI \qquad \text{(Equation (3.6))}$$

it is clear that resistances can be calculated either by assuming a current and working out the potential difference, or by the opposite process. It is also possible to make use of energy methods, with the upper and lower bounds for the resistance being given by these two approaches. In simple cases the resistance of a conductor can be calculated directly.

Worked Example 3.1

A diffused resistor in an integrated circuit is made by diffusing a thin layer of p-type impurity into an n-type isolation island. Assuming that the conductivity of the p-type layer varies linearly from σ_0 at the surface to zero at the interface with the n-type layer, and that the interface is d below the surface, find an expression for the resistance between opposite edges of a square section of p-type layer whose transverse dimensions are much greater than its thickness.

Details of the construction of integrated circuit resistors may be found in Bar-Lev.*

Solution Consider an elementary sheet of the p-type layer of thickness dz lying z below the surface, as shown in Fig. 3.4. Because we shall want to add up the resistances of all such sheets in parallel, it is best to use the conductance $G = 1/R$. The conductivity of the sheet is given by $\sigma = \sigma_0(1 - z/d)$ so the conductance of the sheet is

$$\mathrm{d}G = \sigma_0 (1 - z/d)(L\,\mathrm{d}z/L)$$

The conductance of the whole layer is thus

$$G = \sigma_0 \int_0^d (1 - z/d)\,\mathrm{d}z = \sigma_0 d/2$$

so the resistance in ohms is

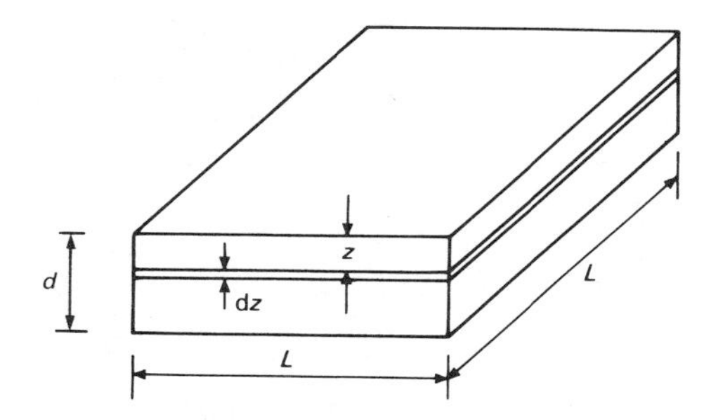

Fig. 3.4 Calculation of the sheet resistance of an integrated-circuit diffused resistor.

$$R = 1/G = 2/(\sigma_0 d)$$

Note that the resistance is independent of the size of the square chosen. This is a general result applying to any sheet of resistive material. For this reason the resistances of resistive sheets are usually quoted in 'ohms per square' (ohms/□). In this example we have, in effect, assumed unit potential difference across the square and calculated the sum of the currents flowing in the elementary sheets.

The approach illustrated by the preceding worked example is only of use in the relatively small number of cases where the shape of the conductor is such that either the current or the potential distribution is simple enough for the integrals to be evaluated. This is clearly not the case with shapes such as that shown in Fig. 3.5. Problems of this kind can be solved by the following procedure:

1. Assume a value for the potential difference V between the ends of the strip
2. Solve Laplace's equation with the appropriate boundary conditions
3. Calculate the electric field and current density distributions on any convenient line across the strip using $\boldsymbol{E} = -\nabla V$ and $\boldsymbol{J} = \sigma \boldsymbol{E}$.
4. Integrate $\boldsymbol{J}$ across the strip to find the total current I.
5. $R = V/I$.

These steps can be carried out by numerical methods if necessary.

This method can sometimes be a bit tricky to apply because of the need to ensure that the normal component of $\boldsymbol{E}$ is zero at the edge of the strip. We can sometimes get round the difficulty by using the **principle of duality**. In a field map such as Fig. 3.5 the field lines and the equipotentials are always at right angles to each other. We can therefore imagine a second problem whose field pattern is identical except that the roles of the field lines and the equipotentials have been exchanged. Two problems related to each other in this way are called *duals* of each other.

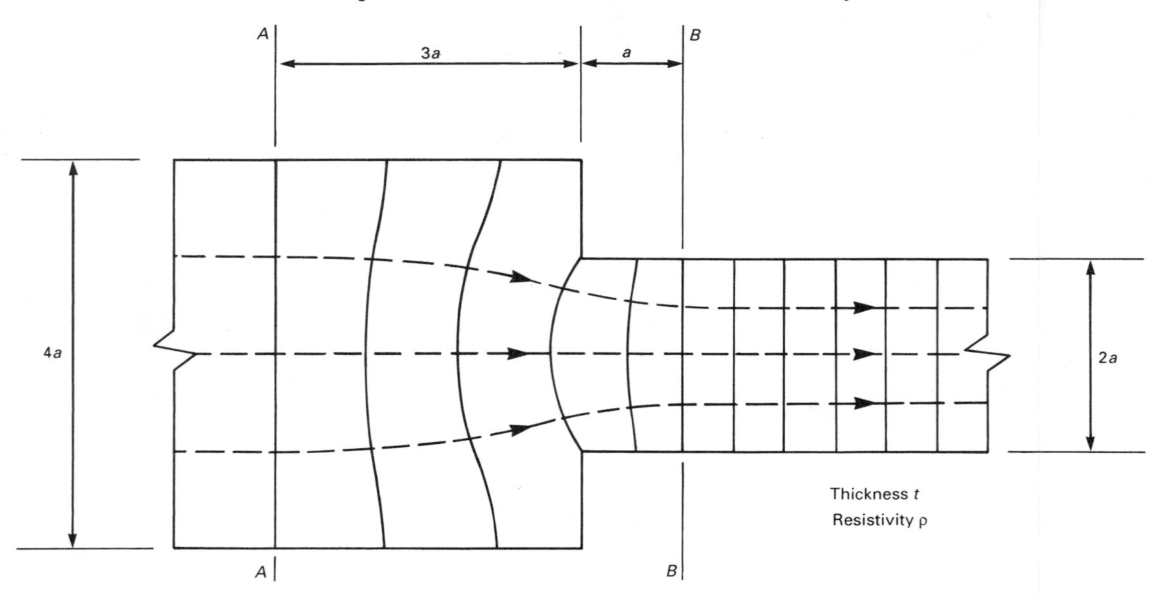

Fig. 3.5 Electric field lines and equipotential surfaces for a current flowing through a conducting strip with a step change in width.

Sometimes it is easier to solve the dual of a problem and then to deduce the required answer from that solution.

Find the resistance between AA and BB of the conducting strip shown in Fig. 3.5.

Worked Example 3.2

Solution Sufficiently far from the step the current densities in the strip may be assumed to be uniform. Figure 3.5 suggests that this will be so for distances from the step exceeding $3a$ to the left and a to the right. Let us therefore assume that at and beyond these distances the current flow is uniform and parallel to the axis. The step in the width of the strip makes the numerical solution of the problem a bit tricky. So, instead, we consider the dual problem, in which the lines of electric field are directed across the strip. The symmetry of the problem means that the whole solution can be calculated if the part on one side of the centre line is known.

Figure 3.6 shows the potentials at the mesh points for the finite difference solution of the dual problem. In terms of the original problem, these figures represent the percentage of the total current flowing between a mesh point and its mirror image on the other side of the axis. Thus the current density at R is approximately $0.132I/(0.5at)$, where I is the total current in the strip. If the resistivity of the strip is ρ, then the electric field on the axis at R is given approximately by

$$E = \rho J = 0.264\rho I/at$$

The potential difference between P and Q is obtained by using the trapezoidal rule to evaluate $\int_P^Q \mathbf{E}\cdot d\mathbf{l}$ to give $V_{PQ} = 1.39\rho I/t$. The resistance between AA and BB is therefore $1.39\rho/t$.

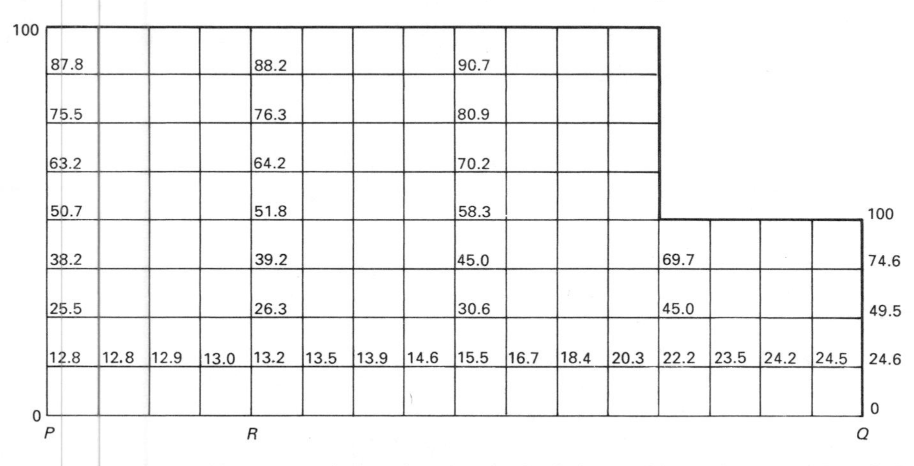

Fig. 3.6 Finite difference solution for the dual of the problem shown in Fig. 3.5.

Calculation of resistance by energy methods

It has already been observed that capacitances can be calculated by energy methods to an accuracy which is adequate for many purposes. Similar methods can be used to calculate resistances. In this case it is assumed that the effect of

The authoritative text on the use of energy methods is Hammond.*

perturbing the distribution of the electric field and the current density from equilibrium is to increase the rate of energy dissipation. Now the power dissipated in a resistance can be written as

$$W = I^2R = V^2/R$$

so, if the total current is given, and the power dissipated is calculated as W′ from an approximate distribution of the current, then

$$W' \geqslant I^2R \qquad \text{and} \qquad R \leqslant W'/I^2 \tag{3.19}$$

giving an upper bound for the resistance. If, on the other hand, the potential difference across the resistance is given and the power is W'' from an approximate set of equipotentials

$$W'' \geqslant V^2/R \qquad \text{and} \qquad R \geqslant V^2/W'' \tag{3.20}$$

giving a lower bound for R. To illustrate how the method works let us consider the problem already solved by numerical methods.

Worked Example 3.3 Use energy methods to obtain an approximate value for the resistance of the conducting strip shown in Fig. 3.5.

Solution To obtain the upper bound for R we must assume some pattern of current flow. Figure 3.7 shows one possibility: no current flows at all in the shaded area of the diagram and the total current is assumed to be uniformly distributed across any plane perpendicular to the axis. It is convenient to divide the problem into two parts as shown and to calculate the resistances R_1 and R_2 separately. If the total current is I, then the power dissipated in a strip of width dx at the plane x is

$$\mathrm{d}W' = \frac{3\rho I^2\,\mathrm{d}x}{2t\,(6a - x)}$$

Integrating from $x = 0$ to $x = 3a$ gives the total power dissipated in the left-hand part of the strip

$$W' = (3\rho I^2/2t)\ \ln 2$$

so that

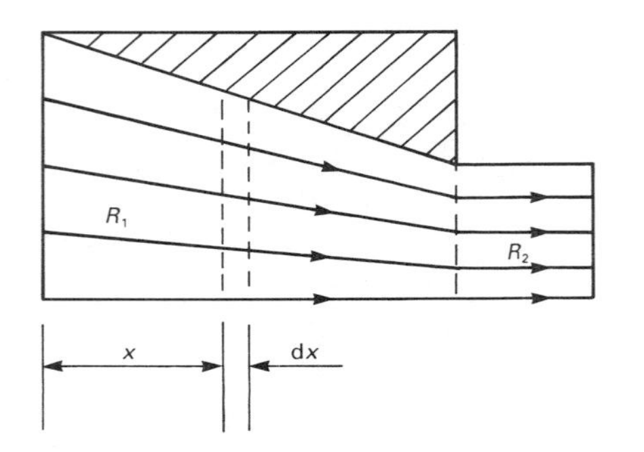

Fig. 3.7 Approximate current flow lines to find an upper bound for the resistance of the strip shown in Fig. 3.5.

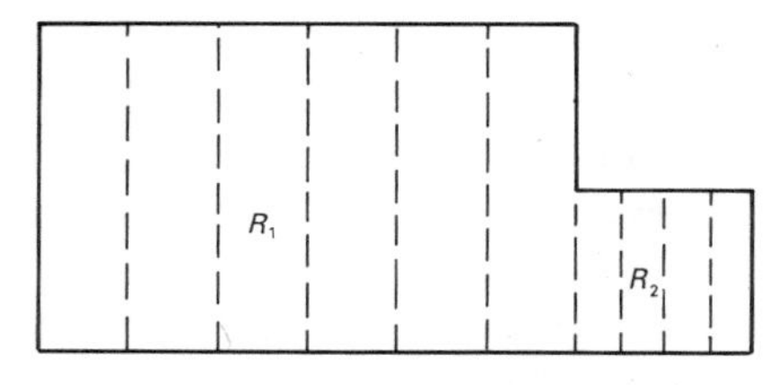

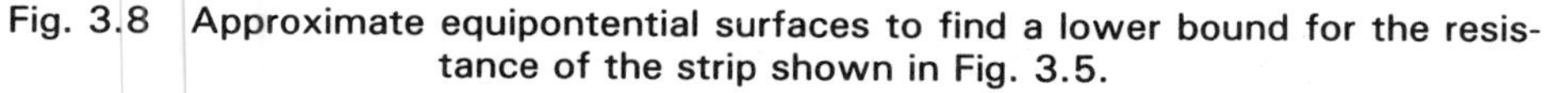

Fig. 3.8 Approximate equipontential surfaces to find a lower bound for the resistance of the strip shown in Fig. 3.5.

$$R_1 \leqslant (3\rho/2t)\ln 2 = 1.04\rho/t$$

The resistance of the right-hand part of the strip is found by elementary methods to be

$$R_2 = 0.5\rho/t$$

Thus the upper bound of the resistance of the strip is

$$R \leqslant 1.54\rho/t$$

To obtain the lower bound to R we assume that the equipotential surfaces are planes perpendicular to the axis, as shown in Fig. 3.8. The resistances R_1 and R_2 can be calculated by elementary methods:

$$R_1 = 0.75\rho/t \qquad \text{and} \qquad R_2 = 0.5\rho/t$$

giving the lower bound $R \geqslant 1.25\rho/t$. The resistance of the strip is therefore

$$R = 1.40\rho/t \pm 11\%$$

which is very close to the result obtained with greater effort by finite difference methods.

Summary

In this chapter we have explored the links between the theory of electromagnetism and that of electric circuits. It has been shown that Ohm's law and Kirchhoff's laws are special cases which arise when electromagnetic theory is applied to lumped components and to electrical networks. On occasion it is necessary to calculate the resistance of a piece of conducting material. Three possible approaches have been discussed: direct calculation, the use of Laplace's equation and the use of energy methods. Direct calculation is possible only in a limited number of cases where the current density is uniform over surfaces of simple shape. The application of Laplace's equation can, in principle, solve all problems involving uniform conducting materials. Similar numerical methods can be used when the materials are not uniform. These methods are available as computer packages of great sophistication. Their disadvantage is the considerable labour involved in using them even when computer packages are employed. For many purposes great accuracy is not required and then the energy methods are of value because of their ability to provide acceptable accuracy with little effort.

Problems

3.1 A current of 1 A flows in a copper wire 1 mm in diameter. Given that there are about 10^{29} conduction electrons per cubic metre in copper, calculate their drift velocity. If the current is the rms value of a 50 Hz alternating current, how far do the electrons move along the wire? What is the power dissipated per metre in the wire? ($\rho = 17.7 \times 10^{-9}\,\Omega\,\text{m}$)

3.2 A coaxial cable has insulation made of a slightly conducting material of resistivity ρ. Given that the inner and outer radii of the insulation are a and b, respectively, find an expression for the leakage resistance per unit length between the inner and outer conductors.

The theory of the pinch resistor is essentially the same as that of the junction field effect transistor (JFET) below pinchoff. A full discussion is given by Bar-Lev.* Note that V is not the terminal voltage of the device because there is a built-in potential difference of about 1 V between the p and n regions.

3.3 Figure 3.9 shows the arrangement of a pinch resistor in an integrated circuit. If the p regions are earthed and the n region is always positive, the two p–n junctions are reverse biased. A layer of the n region adjacent to each junction is depleted of conduction electrons and is, effectively, an insulator. The thickness of the depletion layer is given by $t = a\sqrt{[(V + V_{GS})/V_p]}$, where V_p is a constant and V is the local potential in the channel referred to S. Given that the n channel has width w and its other dimensions are as shown in Fig. 3.9, find expressions for the current in the channel:

(a) when the current through the resistor is small, so that the voltage V_{DS} is much less than V_{GS}; and

(b) when the current through the resistor is not small.

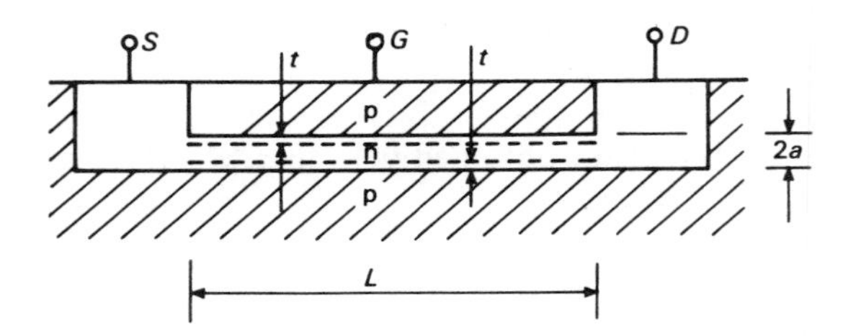

Fig. 3.9 Arrangement of a pinch resistor for an integrated circuit.

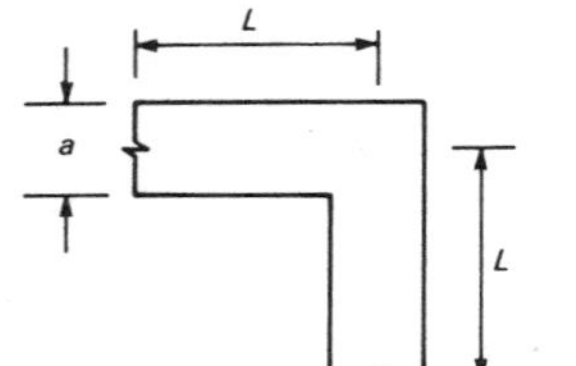

3.4 The figure in the margin shows a right-angled corner in a conducting bar of square section. Estimate the length of straight bar which would have the same resistance as the corner.

3.5 A burglar alarm system works by detecting the change in the resistance of a thin conducting film on a window when the window is broken. A window is typically 1 m square, and contact with the film is made by conducting strips at the top and bottom of the window. Estimate the percentage change in the resistance which would have to be detected by the electronic circuits if a burglar cut a hole just large enough to pass his arm through. What difference does it make where the hole it cut?

The magnetic effects of electric currents

4

Objectives

- ☐ To explain the relationship between magnetism and electricity.
- ☐ To introduce the Biot–Savart law and the magnetic circuit law and to show how they are used to calculate the magnetic flux density produced by electric currents.
- ☐ To introduce the concept of magnetic scalar potential and to discuss the types of problem in which it can be used.
- ☐ To show how problems involving the motion of charged particles in electric and magnetic fields can be solved.
- ☐ To show that forces are exerted on current-carrying conductors in magnetic fields.

The law of force between two moving charges

The theory of magnetism was developed independently from that of electricity until Oersted (1820) showed that the two subjects were linked. This link was later explored by Ampère, Faraday and Maxwell, but it was not until the theory of relativity was developed by Einstein that the relationship between them could be fully understood. It is now clear that magnetic effects can be regarded as a consequence of the motion of electric charges. It is not necessary to postulate the existence of magnetic poles or dipoles to explain experimental observations. The magnetic properties of materials, discussed in Chapter 5, can be explained by assuming the existence of circulating currents on the atomic scale. In this chapter we shall confine our attention to magnetic effects in free space.

The figure in the margin shows two charges which have velocities $\boldsymbol{v}_1$ and $\boldsymbol{v}_2$. It can be shown that the force exerted on Q_2 by Q_1 is given by

$$\begin{aligned} \boldsymbol{F} &= \frac{Q_1 Q_2}{4\pi\epsilon_0 r^2}\hat{\boldsymbol{r}} + \frac{\mu_0 Q_1 Q_2}{4\pi r^2}(\boldsymbol{v}_2 \wedge (\boldsymbol{v}_1 \wedge \hat{\boldsymbol{r}})) \\ &= \boldsymbol{F}_e + \boldsymbol{F}_m \end{aligned} \tag{4.1}$$

The law of force between two moving charges can be derived from the inverse square law of electrostatics using the theory of relativity. Equation (4.1) is the limit of this formula for velocities which are small compared with the velocity of light. The full equation and its derivation are given by Lorrain, P. and Corson, D.R., *Electromagnetic Fields and Waves*, Freeman (1970).

where $\hat{\boldsymbol{r}}$ is a unit vector pointing from 1 to 2 and the symbol $\wedge$ represents the vector cross product. The first term in this equation is the electrostatic force (Equation (1.1)). The second term, which we shall recognize as the magnetic force, is a result of the motion of the particles. The constant μ_0 is known as the **primary magnetic constant**; it has the value $4\pi \times 10^{-7}\,\mathrm{H\,m^{-1}}$. It is instructive to compare the magnitudes of the two forces F_e and F_m.

To simplify matters let us suppose that $\boldsymbol{v}_1$ is parallel to $\boldsymbol{v}_2$. This gets rid of the vector products without affecting the order of magnitude of the result. The ratio of the magnitude of the magnetic force to that of the electrostatic force is then

$$\frac{F_m}{F_e} = \epsilon_0 \mu_0 v_1 v_2 = \frac{v_1 v_2}{c^2} \tag{4.2}$$

The values of ϵ_0 and μ_0 can be established by measurements at low frequencies. Their product is equal to $1/c^2$ within the limits of experimental error. It is shown in Chapter 8 that c is the velocity of propagation of electromagnetic waves.

It can be shown that the constant c is the velocity of light, so c^2 has a numerical value which is approximately $10^{17}\ \mathrm{m^2\,s^{-2}}$. Thus the magnetic force is negligible compared with the electrostatic force unless either:

1. both particles have velocities close to that of light (which happens in some high-power electron tubes) or
2. the electrostatic component of the force is cancelled by the electrostatic force of an equal and opposite stationary charge. The latter is just what happens when electrons are moving along conducting wires because those contain equal numbers of fixed positive charges. Although the electrons move at velocities much less than that of light, there are enormous numbers of them, so useful forces are produced. That is why magnetic forces can be used to generate electric power and drive electric motors. Note that the cross product in Equation (4.1) means that the magnetic force on Q_2 is always at right angles to $\boldsymbol{v}_2$.

The magnetic flux density

In the development of the theory of electrostatics in Chapter 1 the sources of electrostatic force were separated from their effects by introducing the concept of the electric field. Following the same approach, we introduce the vector $\boldsymbol{B}$ and split the magnetic part F_m in Equation (4.1) into two parts:

$$\boldsymbol{B} = \frac{\mu_0 Q_1}{4\pi r^2} (\boldsymbol{v}_1 \wedge \hat{\boldsymbol{r}}) \tag{4.3}$$

and

$$\boldsymbol{F} = Q_2(\boldsymbol{v}_2 \wedge \boldsymbol{B}) \tag{4.4}$$

The older c.g.s. unit of magnetic flux density, the gauss (G), is still in common usage. $1\ \mathrm{T} = 10^4\ \mathrm{G}$.

The vector $\boldsymbol{B}$ is referred to, somewhat confusingly, as the *magnetic flux density*. It is measured in units known as either **teslas** (T) or webers (Wb) per square metre. Equation (4.3) provides the basis for calculating the magnetic flux density produced by any combination of charges moving in free space. The effect of that flux density on any other charges can then be found by using Equation (4.4). For the moment we will concentrate on ways of computing magnetic flux densities. This information might be needed to calculate:

- the forces on moving charges;
- the forces on current-carrying conductors;
- the forces on iron surfaces;
- self- and mutual inductances;
- induced voltages and currents;
- eddy current losses.

The second and third of these are mainly applied to electric machines, and we shall refer to them only briefly.

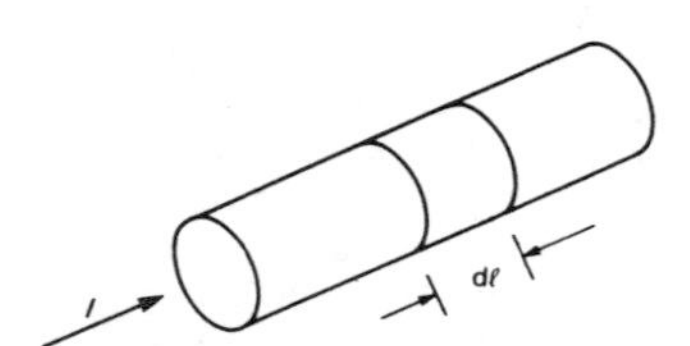

The source of the magnetic flux is often a current flowing in a wire, so another form of Equation (4.3) is needed. Consider an element of length dl in a wire carrying current I as shown in the figure in the margin. Let the current consist of charges of magnitude q moving with mean drift velocity v_d. If there are n charge carriers per unit volume and the cross-sectional area of the wire is A, then the current in the wire is given by $I = n\,q\,A\,v_d$, from Equation (3.2). But the charge contained in the element of wire is $nqA\,dl$, so that

$$Qv = (n\,q\,A\,\mathrm{d}l)\,v_{\mathrm{d}} = I\,\mathrm{d}l \tag{4.5}$$

The contribution to the magnetic flux density from the charges within the element is therefore, from Equation (4.3),

$$\mathbf{d}\boldsymbol{B} = \frac{\mu_0(nqA\,\mathrm{d}l)}{4\pi r^2}\,(\boldsymbol{v}_{\mathrm{d}} \wedge \boldsymbol{r}) = \frac{\mu_0 I}{4\pi r^2}\,(\mathbf{d}\boldsymbol{l} \wedge \hat{\boldsymbol{r}}) \tag{4.6}$$

But isolated current elements cannot exist so it is necessary to integrate Equation (4.6) around the complete circuit in which the current is flowing. The result is

$$\boldsymbol{B} = \frac{\mu_0 I}{4\pi}\oint \frac{\mathbf{d}\boldsymbol{l} \wedge \hat{\boldsymbol{r}}}{r^2} \tag{4.7}$$

This equation is known as the **Biot–Savart law**. It is more straightforward to apply than might appear at first sight, as the following example will show.

Worked Example 4.1

Figure 4.1 shows a single circular loop of wire carrying current I. Find an expression for the magnetic flux density at the point P on the axis of the loop.

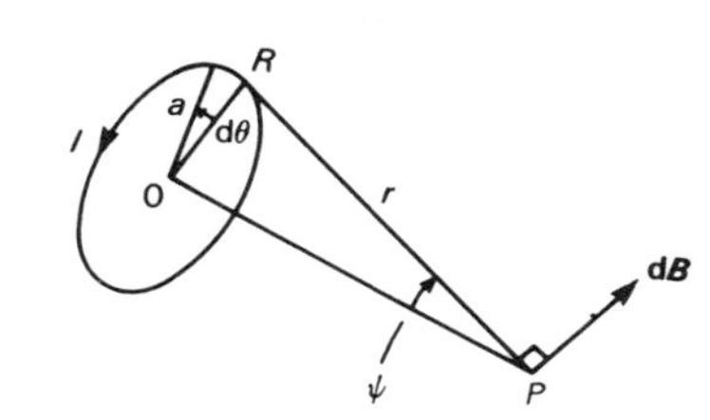

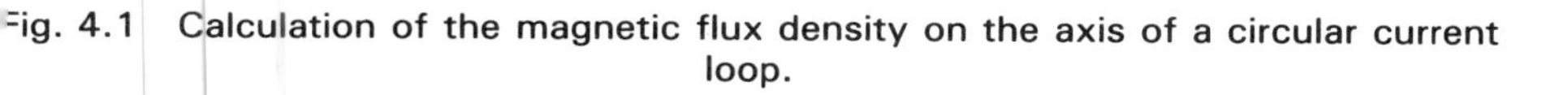

Fig. 4.1 Calculation of the magnetic flux density on the axis of a circular current loop.

Solution Consider the flux density at P due to the current element at R. The current element has length $a\,\mathrm{d}\theta$ and its distance from P along the line RP is r. From the triangle ORP, $r = a/\sin\psi$. The flux density produced by the current element is represented by the vector $\mathbf{d}\boldsymbol{B}$, which must be normal to both RP and the current element according to the rule for vector products. The vector $\mathbf{d}\boldsymbol{B}$, therefore, lies in the plane ORP at right angles to RP, as shown in Fig. 4.1. Because RP is at right angles to the current element it follows from Equation (4.6) that

$$\mathrm{d}B = \frac{\mu_0 I}{4\pi a}\sin^2\psi\,\mathrm{d}\theta$$

This can be resolved into its axial and radial components

$$\mathrm{d}B_{\mathrm{r}} = \mathrm{d}B\sin\psi \qquad \text{and} \qquad \mathrm{d}B_{\mathrm{z}} = \mathrm{d}B\cos\psi$$

Integrating around the loop we find that

surface S which spans the closed curve. This equation, like Gauss' theorem in electrostatics, can be applied usefully to problems only when it is possible to guess the distribution of the magnetic flux density. In free space this restricts its use to those few cases which have obvious symmetry. We shall see in Chapter 5 that it has a very important use in the solution of problems in which the greater part of the magnetic flux is carried by an iron circuit.

Worked Example 4.3 Figure 4.4 shows a toroidal coil having an inner radius R_1 an outer radius R_2 and N turns of wire carrying current I. Find the magnetic flux density at any point.

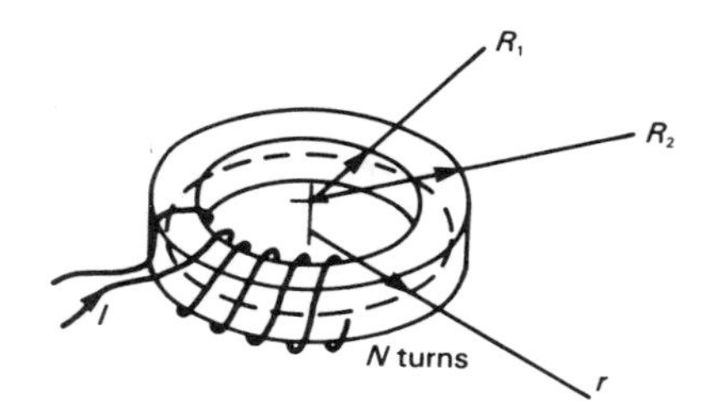

Fig 4.4 The arrangement of a toroidal coil.

Solution From consideration of symmetry, the flux density within the winding must be in the tangential direction and vary only with radius. Applying the magnetic circuit law to the circular path of radius r shown by the broken line in the diagram gives

$$\oint \mathbf{B} \cdot \mathbf{d}\mathbf{l} = 2\pi r B$$

while the current encircled is NI. Hence $B = NI/2\pi r$ within the winding. Any surface spanning a path of integration lying outside the winding is crossed by equal numbers of conductors carrying currents in each direction so that the magnetic flux density outside the toroid is apparently zero. More careful thought shows that if the winding is a single layer, then current I must encircle the toroid once in the tangential direction. The magnetic flux density is therefore given by Equation (4.8) at points outside the toroid and remote from it. If it is necessary to produce a toroid with no flux outside it then a winding with an even number of layers is used. Each layer is wound in the opposite direction around the toroid from the one below it so that the tangential component of the current is cancelled.

Magnetic scalar potential

The theory of magnetism was originally developed in parallel with that of electrostatics, with magnetic poles as the equivalent of electric charges. This approach is to be found in all the older textbooks.

Although the theory of magnetism has been developed here without using the idea of magnetic poles, it is useful to introduce the magnetic analogue of electric potential. This concept is a convenient mathematical device which does not have any physical significance. The magnetic **scalar potential** between the points P and Q is given by

$$\mathscr{V}_Q - \mathscr{V}_P = -\frac{1}{\mu_0}\int_P^Q \boldsymbol{B}\cdot\mathbf{d}\boldsymbol{l} \tag{4.13}$$

Although the idea of magnetic pole strength is still occasionally useful it has the disadvantage that it obscures the relationship between magnetism and electricity.

where the constant μ_0 has been introduced so that Equation (4.13) has the same form as the equations given in texts which assume the existence of magnetic poles. The magnetic scalar potential is of limited usefulness because it can be given a unique value only in a region of space in which there are no electric currents. This point is illustrated by the figure in the margin. It can be shown from Equation (4.10) that the difference in the scalar potential between P and Q is independent of the path of integration chosen provided that it does not encircle any electric current. If, however, a path such as 2 were chosen for the integration, the result would be to subtract I from the value of the potential difference calculated.

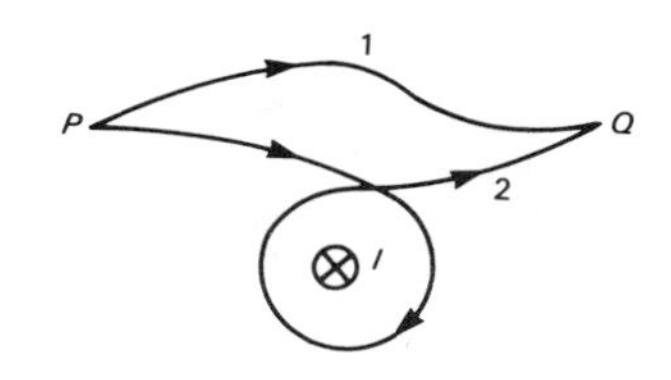

What is worse still, is that by making several loops around the current it is possible to make the potential difference take a whole series of different values, because the value of the integral changes by I for each complete loop made around the current. Thus the magnetic scalar potential can be given a unique value only in a region of space containing no currents. Despite this limitation there are still times when the idea is very useful. The differential form of Equation (4.13) is

$$\boldsymbol{B} = -\mu_0 \,\mathrm{grad}\, \mathscr{V} = -\mu_0 \nabla \mathscr{V} \tag{4.14}$$

Moreover, since it has been assumed that free magnetic poles do not exist, the magnetic analogue of Gauss' theorem is

The integral in Equation (4.15) is taken over a closed surface.

$$\oiint \boldsymbol{B}\cdot\mathbf{d}\boldsymbol{S} = 0 \tag{4.15}$$

or, in words, the *flux of* ***B*** *out of a closed surface is zero*. The unit of magnetic flux is the **weber** (Wb). The differential form of Equation (4.15) is

$$\mathrm{div}\,\boldsymbol{B} = \nabla\cdot\boldsymbol{B} = 0 \tag{4.16}$$

Combining Equations (4.14) and (4.16), we find that $\mathscr{V}$ must obey Laplace's equation in free space just as the electric potential does. That is

$$\mathrm{div\,grad}\,\mathscr{V} = \nabla^2\mathscr{V} = 0 \tag{4.17}$$

The commonest use of the magnetic scalar potential is for calculating the magnetic flux density in the space between iron pole pieces. We shall return to this subject in Chapter 5.

Forces on charges moving in magnetic fields

We now return to Equation (4.4), the second of the two parts into which the law of magnetic force was split, and consider the motion of charged particles in magnetic fields. The magnetic force on a moving charged particle is put to use in deflecting the electron beam in a television picture tube, in controlling the motion of electrons in microwave tubes, in focusing the electron beam in an electron microscope, and in many other devices. The force on a moving charge is given by

The accident of history which made the sign of the charge on the electron negative provides many pitfalls for the unwary. I prefer to work as though q were a positive charge and then sort out the correct sign for the direction of the force at the end of the calculation using physical arguments.

$$\boldsymbol{F} = q(\boldsymbol{v}\wedge\boldsymbol{B}) \tag{4.4}$$

This force always acts at right angles to the direction of motion of the particle and therefore **does no work on it.** The magnetic force acting on a particle can therefore change its direction of motion but not its speed. The simplest case is that of a particle moving at right angles to the direction of a uniform magnetic field, as shown in the figure in the margin. Because the particle is acted on by a constant force which is perpendicular to its direction of motion, it must move in a circle, as shown. The radial acceleration is provided by the magnetic force, so that

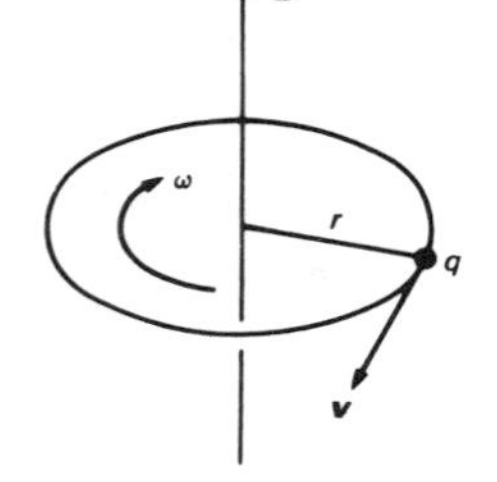

$$mr\omega^2 = qvB = qr\omega B$$

whence

$$\omega = \frac{q}{m}B \tag{4.18}$$

This frequency is known as the cyclotron frequency, from its application in the particle accelerator of that name. For an electron in a magnetic flux density of 0.1 Tesla the cyclotron frequency is about 2.8 GHz. If the particle has an initial component of velocity parallel to the vector $\boldsymbol{B}$, then that component is unaffected by the magnetic field. The motion of the particle is then helical from the combination of the steady motion in the direction of $\boldsymbol{B}$ with the circular motion in the direction perpendicular to it.

1 GHz = 10^9 H. This frequency is at the lower end of the microwave part of the spectrum of electromagnetism waves.

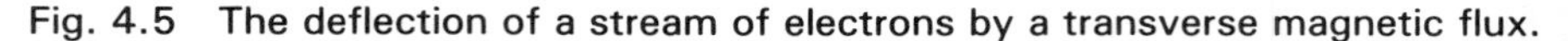

Worked Example 4.4

An electron moving with velocity v enters a region of space in which the magnetic flux density is B at right angles to the velocity of the electron, as shown in Fig. 4.5. Given that the magnetic field is constant for a distance L in the direction of the initial motion of the electron and then falls abruptly to zero, find an expression for the angular deflection of the electron motion by the magnetic field.

This example is obviously an oversimplification because the flux density would not fall abruptly to zero at the edges of the region. The solution of the same problem with a realistic distribution of the flux density would probably require numerical methods, although it would introduce no new physical principles.

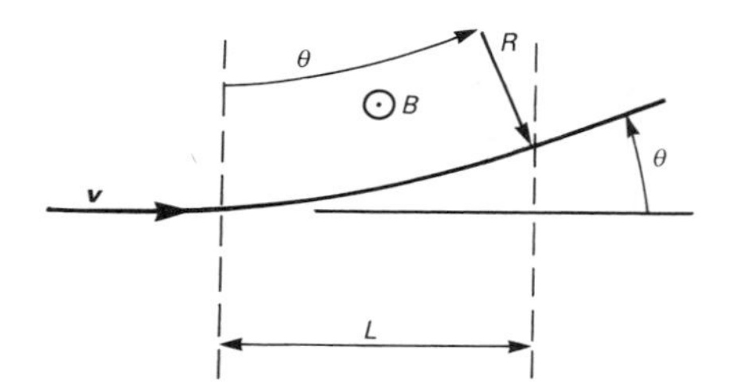

Fig. 4.5 The deflection of a stream of electrons by a transverse magnetic flux.

Solution The electron must move along a circular path of radius R, as shown, such that

$$mv^2/R = qvB$$

so

$$R = mv/qB$$

But

$$R\sin\theta = L$$

so the angular deflection is given by

$$\theta = \arcsin\ (qBL/mv)$$

This type of deflection system is used in television picture tubes. In practice the field is not constant. It is provided by a set of saddle coils rather like those shown in Fig. 4.6 but modifed to fit around the conical neck of the tube. The system is capable of producing very wide-angle deflections (around 55°) but is relatively slow to respond to voltage waveforms because of the inductance of the scanning coils. For large angles of deflection the deflection of the spot on the screen is not proportional to B, and specially compensated driving circuits are used.

Motion of charges in combined electric and magnetic fields

When the motion of a charged particle is controlled by combined electric and magnetic fields the result can be very complex. There is not room in a book of this length to discuss this type of problem in detail, so you should consult more specialized texts for further information. One example which is of considerable practical importance is given below as a worked example, and some more may be found among the problems at the end of the chapter.

For a detailed discussion of problems of this type consult Spangenburg.*

Worked Example 4.5

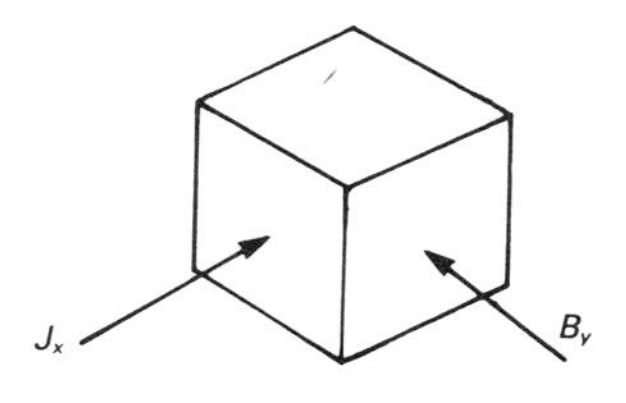

An electric current of density J_x flows through a sample of semiconductor material in the x direction. A uniform magnetic field B_y is imposed in by y direction and the resulting deflection of the charge carriers produces a surface charge on the sample and an electric field E_z in the z-direction which balances the magnetic force on the charges. Find an expression for the electric field strength in terms of J_x, B_y, the charge of each charge carrier and the density of charge carriers.

Solution In equilibrium no current flows in the z-direction and the net force on a charge carrier in the z-direction must be zero. Therefore

$$qE_z + qv_xB_y = 0$$

where q is the charge on a charge carrier (assumed positive) and v_x is the mean drift velocity of the carriers. Now $J_x = nqv_x$, where n is the number density of the carriers, so

$$E_z = -v_xB_y = -J_xB_y/nq$$

The electric field E_z is associated with a potential difference between the faces of the sample which can be measured by a high-impedance voltmeter. As the product J_xB_y is positive, the magnitude of the voltage is a measure of the density of charge carriers in the sample and its sign shows their sign. This effect, known as the **Hall effect**, is used routinely to measure the properties of samples of semiconductor material. It is also used as a means of measuring magnetic flux density.

For details of the application of the Hall effect in measuring the properties of semiconductor materials, see Bar-Lev.* Compton* discusses some other applications.

Ferromagnetic materials

The overwhelming majority of materials which exist in the world have magnetic properties which differ so little from those of a vacuum that, for nearly all engineering purposes, they are indistinguishable from it. The exceptions to this rule are the elements iron, cobalt and nickel, together with certain materials which include them. Their special properties are illustrated by Fig. 5.1. Figure 5.1(a) shows an air-cored solenoid carrying a current I. The flux density in the vicinity of the solenoid, somewhat as shown, could be mapped with the aid of a flux meter. It is worth emphasizing that there would be no measurable change in the field if the solenoid were filled with a core made of brass, aluminium, wood, plastic, or any of the non-ferromagnetic materials. If, on the other hand, an iron core were inserted the result would be a considerable increase in the flux density, as shown in Fig. 5.1(b).

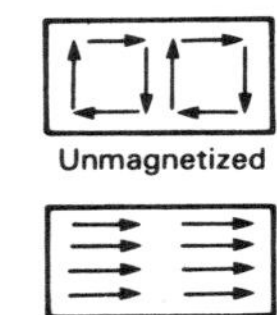

The physical explanation of ferromagnetism is to be found in books for physicists such as Bleaney and Bleaney.*

This property of iron arises because the iron atoms are themselves tiny permanent sources of magnetic flux like current loops. If each of these 'loops' is represented by an arrow to show its polarity, then a qualitative explanation of ferromagnetism can be given. In an unmagnetized piece of iron the atomic magnets are arranged, head to tail, in closed loops. All the flux then circulates within the iron, none of it outside. The effect of applying an external flux density to the iron is to cause the atomic magnets to line up with the external flux so that their flux is added to it. These two situations are illustrated diagrammatically in the figures in the margin. The actual processes of magnetization are more complex than this, but there is no room in this book to enter into them in more detail. The macroscopic phenomena of ferromagnetism are discussed later in this chapter.

From Fig. 5.1(b) we observe that the field outside the solenoid could just as well be produced by the addition of a second winding, with a suitable current and distribution of turns, to the air-cored solenoid of Fig. 5.1(a). This suggests that, just as the polarization of dielectric materials can be represented by a surface-charge distribution, so the polarization of ferromagnetic materials can be repre-

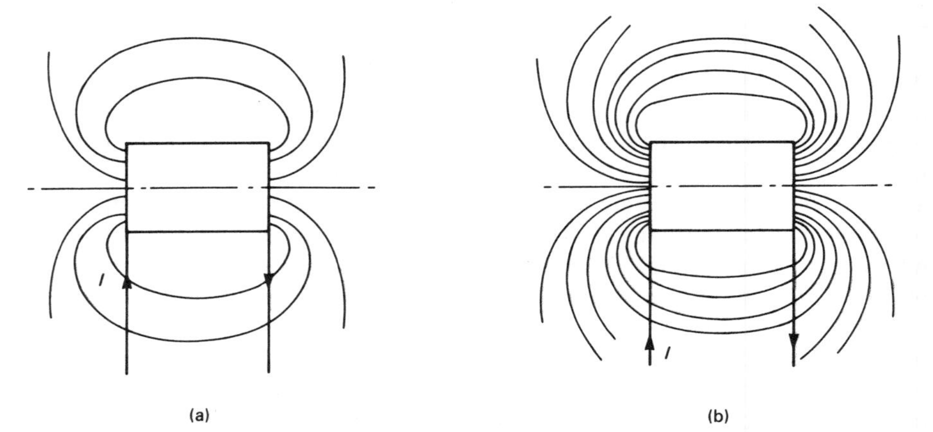

Fig. 5.1 The magnetic fields of (a) an air-cored solenoid and (b) a similar solenoid with an iron core, showing diagrammatically the great increase in the strength of the field in the second case.

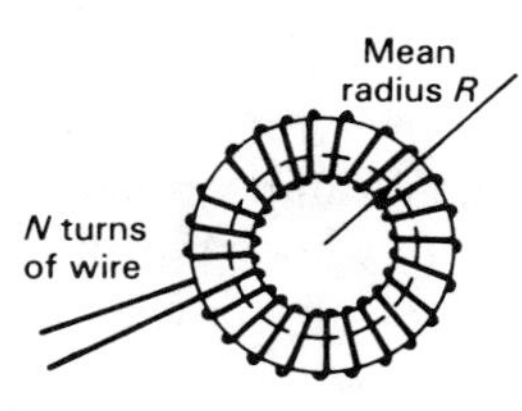

Fig. 5.2 An iron ring of mean radius R with N turns of wire wound uniformly upon it.

sented by a surface-current distribution. This is consistent with the view adopted in this book that magnetic phenomena are best regarded as resulting from the motion of electric charges.

To illustrate the effect of the magnetization of iron we consider an iron ring with N turns of wire wound uniformly upon it as shown in Fig. 5.2. Let the current in the winding be I, and let the magnetization of the iron be represented by an additional current I' in the winding. Then, by the magnetic circuit law:

$$2\pi RB = \mu_0 N(I + I') \tag{5.1}$$

or

$$B = \frac{N}{2\pi R}\mu_0(I + I') \tag{5.2}$$

Now experiments with electromagnets show that the magnetic flux density produced increases with the current, though not necessarily in a linear fashion, so let us assume that $I' = \chi_m I$ where χ_m is the **magnetic susceptibility**.

Then

$$B = \frac{N}{2\pi R}\mu_0 I(1 + \chi_m) \tag{5.3}$$

or, if $\mu_r = 1 + \chi_m$,

$$B = \frac{N}{2\pi R}\mu_0\mu_r I = \frac{N}{2\pi R}\mu I \tag{5.4}$$

where the quantities μ and μ_r are known as the **permeability** and the **relative permeability** of the iron, respectively. The analogy with the case of dielectric permittivity is only superficial because it is hardly ever valid to regard the magnetic permeability as a constant. The other distinguishing feature is that the relative permeability of iron is much greater than the relative permittivity of common materials, being of the order of 10^4.

In problems involving dielectric materials we found it useful to make use of the vector $\boldsymbol{D}$ to avoid having to calculate the distribution of polarization charges. The same approach can be used in magnetic problems. We define a new vector $\boldsymbol{H}$ by the equation

$$\boldsymbol{H} = \boldsymbol{B}/\mu \tag{5.5}$$

so that, for the simple case shown in Fig. 5.2, we have

$$H = NI/2\pi R \tag{5.6}$$

The vector $\boldsymbol{H}$ is known as the **magnetic field**. It is measured in A m^{-1} in SI units,

The fact that it is $\boldsymbol{H}$ which is known as the magnetic field rather than $\boldsymbol{B}$ is a consequence of the historical development of the theory of magnetism in terms of magnetic poles as sources of flux analogous to electric charges.

as can be seen from Equation (5.6). Since the strength of the field is proportional to the product of the current in the winding and the number of turns, we often speak of a field strength of so many 'ampere-turns per metre'. Equation (5.6) also shows that the vector $\boldsymbol{H}$ depends only upon the distribution of electric currents and not upon the arrangement of any iron present.

Making use of the vector $\boldsymbol{H}$, we can deduce a form of the magnetic circuit law (Equation (4.12)) which is valid when magnetic materials are present:

$$\oint \boldsymbol{H} \cdot \mathbf{d}\boldsymbol{l} = \iint \boldsymbol{J} \cdot \mathbf{d}\boldsymbol{A} \tag{5.7}$$

To find the distribution of magnetic flux density in problems involving iron we first calculate $\boldsymbol{H}$ from the known distribution of currents, using Equation (5.7), and then calculate $\boldsymbol{B}$ in each region from Equation (5.5). Magnetic problems cannot usually be solved by the application of potential theory and the use of Laplace's equation because of the non-linearity of magnetic materials. For the same reason it is not possible to make use of the principle of superposition unless linearizing assumptions can be made. This contrasts with the situation in electrostatics where the distribution of charges is frequently unknown but the potentials are given on the conducting boundaries.

Boundary conditions

Whenever a magnetic problem involves two or more regions of space with different magnetic properties it is necessary to use the relationships between the magnetic vectors on the two sides of each boundary.

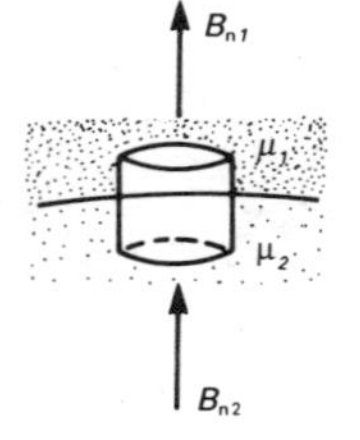

The figure in the margin shows the boundary between two materials having permeabilities μ_1 and μ_2. Consider a Gaussian surface in the form of a 'pillbox' as shown. If the cross-sectional area of the box in the plane of the surface is $\mathrm{d}A$, then the flux of $\boldsymbol{B}$ out of the box is

$$(B_{\mathrm{n1}} - B_{\mathrm{n2}})\,\mathrm{d}A = 0 \tag{5.8}$$

from Equation (4.15), since the contribution of the flux through the sides of the box becomes negligible as the height of the box tends to zero. The consequence of Equation (5.8) may be stated in words: *The normal component of* $\boldsymbol{B}$ *is continuous across a boundary.*

The boundary condition on $\boldsymbol{H}$ is found by applying the magnetic circuit law to a small closed path encircling part of the boundary as shown in the figure in the margin. If the length of the path parallel to the boundary is $\mathrm{d}l$, then the line integral of $\boldsymbol{H}$ around it becomes

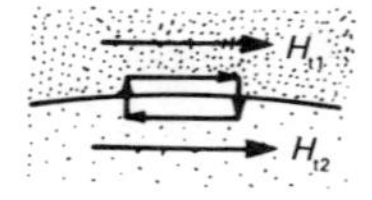

$$(H_{\mathrm{t1}} - H_{\mathrm{t2}})\,\mathrm{d}l = \text{current enclosed} \tag{5.9}$$

It should be noted that the current referred to is an actual electric current. It does not include the hypothetical surface currents which were introduced as a convenient way of describing magnetic materials. In many cases the surface current will be zero and then the boundary condition is that the tangential component of $\boldsymbol{H}$ is continuous across the boundary.

Flux conduction and magnetic screening

Equation (5.9) has one very important practical corollary. Consider an iron bar surrounded by air. Assuming that there is no surface current flowing along the bar, we know that the tangential components of $\boldsymbol{H}$ are equal inside and outside the bar. But, since $\boldsymbol{B} = \mu \boldsymbol{H}$, the component of the magnetic flux density parallel to the bar is much greater inside it than outside it. If such a bar is placed in a magnetic field in which the total flux is fixed, with the axis of the bar aligned with the field, then the requirement that $B_{\text{iron}} \gg B_{\text{air}}$ means that most of the flux will pass through the iron as shown in the figure in the margin. The flux density in the air is then much less than it was before the iron was introduced. The iron therefore acts as quite a good conductor of magnetic flux. The analogy between flux conduction by iron and electric conduction is strengthened still further if we consider what happens to the direction of a flux line entering the iron from the air, as shown in Fig. 5.3. The boundary conditions require that $B_{\text{ni}} = B_{\text{na}}$ and that $B_{\text{ti}} = \mu_r B_{\text{ta}}$. Now μ_r is typically in the range 10^3–10^6, so the flux line must turn through an angle even greater than that indicated in Fig. 5.3. The angle of incidence is greatest when the flux line within the iron lies nearly parallel to the surface. A simple calculation shows that if $\alpha_i = 89°$ and $\mu_r = 1000$, then $\alpha_a = 3.3°$. For many purposes it is adequate to make the approximation that iron is a perfect magnetic conductor, so that all the flux lines enter it at right angles to the surface.

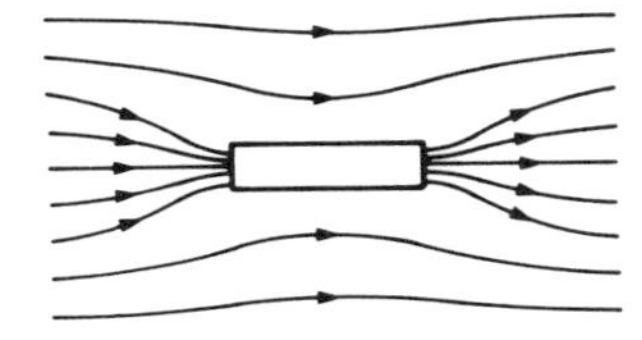

The ability of iron to conduct flux is put to work in a number of ways. It allows magnetic circuits to be constructed which conduct the flux wherever it is required – an idea which we shall consider in detail later on. It also allows flux to be excluded from regions where it is not required. Figure 5.4 shows what happens when a hollow iron cylinder is placed in a magnetic field. The flux lines prefer to pass through the iron and the space within the cylinder is almost entirely free from magnetic flux. This property can be used to shield sensitive equipment from magnetic fields, and special alloys such as mumetal are used for this purpose. The effectiveness of a screen can be expressed in decibels as

Mumetal contains 76% nickel, 17% iron, 5% copper and 2% chromium.

$$S_M = 20 \log_{10}\left(\frac{B \text{ at a point in the absence of the screen}}{B \text{ at the same point with the screen}}\right) \tag{5.10}$$

A mumetal cylinder 100 mm in diameter and 1 mm thick has a magnetic screening effectiveness of 60 dB provided that the external magnetic field is not too strong. It should be noted that it is only low-frequency fields which are screened out by

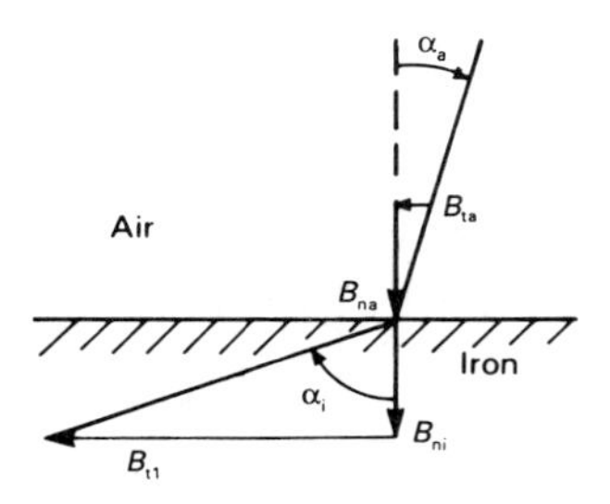

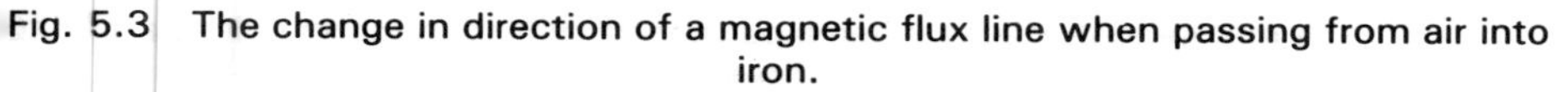
Fig. 5.3 The change in direction of a magnetic flux line when passing from air into iron.

and

$$B_y = -\frac{\mu_0 I}{2\pi}\left(\frac{x}{x^2 + (y+d)^2} + \frac{x}{x^2 + (y-d)^2}\right)$$

Notice that B_x is zero when $y = 0$ as it should be.

For further information consult Hammond, P. Electric and magnetic images, *Proc. I.E.E.*, **107**, Part C (1960), Monograph No. 319.

Strictly speaking, the imaging plane should not quite coincide with the iron surface, because iron is not a perfect conductor of flux, but the difference is not usually important. We shall see later that the flux density in iron is limited by saturation effects. It is possible to regard an iron surface as a magnetic equipotential only if the flux density is below the saturation level.

Magnetic circuits

These assumptions may seem rather sweeping but they do allow us to get answer to an accuracy of 10% or so, which is adequate for many purposes. To get better accuracy would require the use of a large computer.

The analogy between the conduction of electricity by conductors and the conduction of magnetic flux by iron leads us to the useful concept of a magnetic circuit. Figure 5.7 shows a simple magnetic circuit formed by a square iron core with a narrow air gap in it at A. A coil of N turns of wire is wound on the core and carries current I. To make the problem easy to handle we make some simplifying assumptions. Let us suppose that the magnetic field strength is the same everywhere within the iron, having the value H_i, and that the magnetic field in the air gap is H_a. Now consider a closed path around the circuit which follows the centre line (shown dotted). Applying the magnetic circuit law to this path gives

$$4LH_i + gH_a = NI \tag{5.11}$$

Strictly speaking, the length of the iron path is $(4L - g)$ but, provided that g is small, the difference is negligible given the other assumptions which have been made.

The boundary condition (5.8) requires that

$$B_i = B_a = \Phi/A \tag{5.12}$$

where Φ is the total flux circulating, and A is the cross-sectional area of the iron

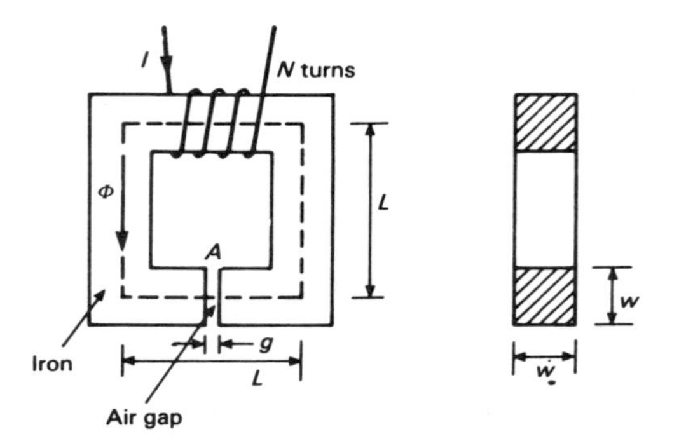

Fig. 5.7 A simple magnetic circuit made up of an iron core with an air gap in it. The flux is supplied by a winding of N turns of wire.

at right angles to the direction of the magnetic field. We have therefore assumed that all the flux due to the coil is contained within the bounds of the iron core and its projection across the air gap. We also know from Equation (5.5) that

$$B_i = \mu H_i \qquad \text{and} \qquad B_a = \mu_0 H_a \tag{5.13}$$

Substitution for H_i and H_a in Equation (5.11) in terms of Φ and the various constants gives

$$NI = \frac{1}{A}\left(\frac{4L}{\mu} + \frac{g}{\mu_0}\right)\Phi \tag{5.14}$$

which can be written

$$\mathscr{M} = \mathscr{R}\Phi \tag{5.15}$$

where $\mathscr{M}$ $(=NI)$ is known as the **magnetomotive force** and $\mathscr{R}$ as the **reluctance**. Equation (5.15) is analogous to Ohm's law for electric circuits, but the analogy must not be pressed too far. Unlike an electric current, the magnetic flux in the circuit dissipates no energy and there are no circulating magnetic charges. Moreover, it cannot be emphasized too often that magnetic materials do not normally behave in a linear fashion, so the reluctance of a circuit can only be regarded as approximately constant at best, and that only under a limited range of conditions. We will return to give this point further consideration after discussing the behaviour of real magnetic materials. First, though, it is useful to mention two other matters: fringing and leakage of magnetic flux.

Fringing and leakage

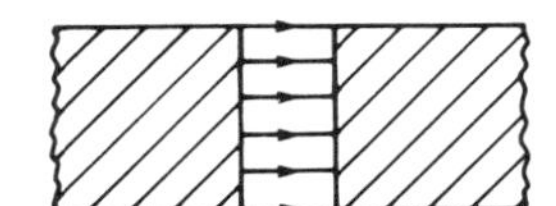

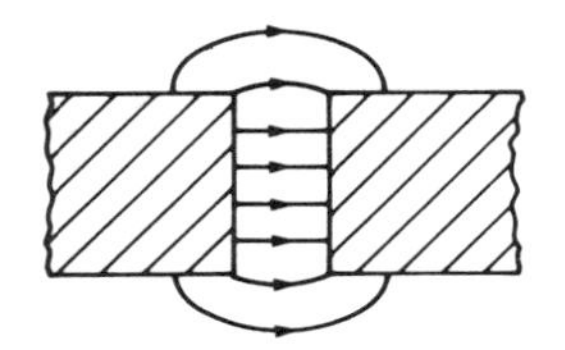

In the discussion of magnetic circuits above it was assumed that the flux lines in the air gap passed straight across, as shown in the upper figure in the margin. A little thought shows that this is an over simplification and that the flux lines must actually spread out into a **fringing field**, as shown in the lower figure in the margin. The truth of this observation can be demonstrated by laying a sheet of paper over the air gap and sprinkling iron filings on it. Clearly the reluctance of the magnetic circuit is affected by the fringing of the field around the gap so we need an estimate of how big the effect is. A detailed solution of the problem would require the solution of Laplace's equation in the air gap (assuming the iron surfaces to be equipotentials). This is unnecessarily complicated for most purposes. A useful rule of thumb is that the effective cross-sectional area of the air gap is larger than that of the pole face by a strip of width $\frac{1}{2}g$ around it, as shown in Fig. 5.8. So Equation (5.12) becomes

$$\Phi = B_i w^2 = B_a (w + g)^2 \tag{5.16}$$

and the reluctance of the circuit is

$$\mathscr{R} = \left(\frac{4L}{\mu w^2} + \frac{g}{\mu_0 (w + g)^2}\right) \tag{5.17}$$

Fringing fields are put to use in magnetic recording heads. Figure 5.9 shows a typical arrangement. Most of the flux in the magnetic circuit passes directly across

The exact solution of the problem just discussed would require a very large computer. Fortunately such high accuracy is not needed in most practical cases. The worked example above shows how figures which are accurate to a few per cent can be obtained by some quite crude approximations. Even the most accurate calculation made using magnetic circuit theory is likely to be in error by a few per cent because of the non-linear properties of the iron which are discussed in the next section.

Hysteresis

For a discussion of the physical reasons for hysteresis in magnetic materials, consult Bleaney and Bleaney.* The word 'hysteresis' comes from the Greek word meaning 'to lag behind'.

In the preceding sections the permeability of iron has been taken to be a constant. While this is a useful first approximation, which can be justified in a limited number of cases, it is not generally possible to regard μ as a constant. The relationship between the magnetomotive force applied to a specimen of iron and the magnetic flux density produced is a far from simple one. The magnetization of the iron depends not just on the present value of the magnetomotive force, but also on the previous history of the specimen. In this book we are restricting our attention to macroscopic phenomena, so no attempt is made to offer an explanation of their causes in terms of the atomic theory of matter.

One way of investigating the relationship between the magnetomotive force applied to a specimen of iron and the flux density produced would be to use a magnetic circuit like the one shown in Fig. 5.7. Provided that the air gap is very small, it is possible to neglect the effects of fringing. The flux density in the air gap is then the same as that in the iron and its strength could be measured by placing a Hall effect probe in the gap. The total magnetomotive force applied to the circuit is NI, so the magnetomotive force applied to the iron is

$$\mathscr{M}_{\text{i}} = NI - \mathscr{M}_{\text{gap}} = NI - Bg/\mu_0$$

Now $\mathscr{M}_{\text{i}}$ is the product of H_{i} and the length of the part of the circuit lying within the iron, so it would be possible to deduce values of H_{i} from measurements of the current in the coil and the flux density in the air gap. The results of such measurements are customarily displayed by plotting B_{i} against H_{i}. There are better ways of obtaining the information than the experiment described, but we are concerned here with the properties of the iron rather than the details of how they are measured.

Figure 5.12 shows a typical B–H curve. If the specimen of iron is initially

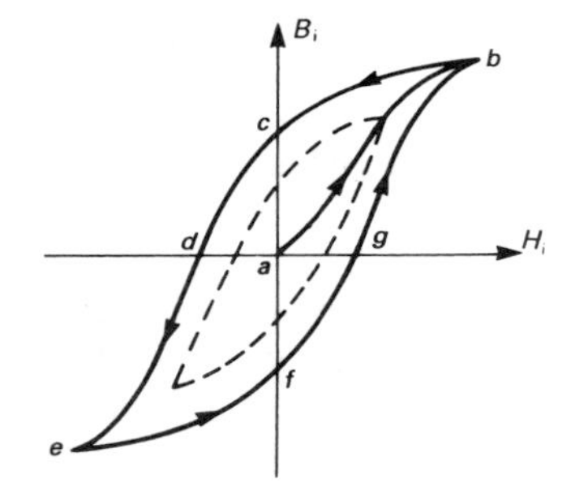

Fig. 5.12 A typical hysteresis loop showing the initial magnetisation path (*ab*), saturation (at *b*), remanence (at *c*), the coercive force (at *d*), and an inner loop (shown dotted).

unmagnetized, then its state can be represented by the point *a* at the origin of the graph. Now suppose that the current in the coil is gradually increased. It is found that the *B–H* curve follows the path *ab*, rising slowly at first, then more rapidly, and, finally, levelling off. This curve is known as the **initial magnetization curve** of the iron. When the curve levels off at high values of H_i the material is said to be **saturated.** If the current in the coil is reduced to zero, the flux density does not retrace the curve *ab*. Instead, it follows a curve such as *bc*, so that the flux density in the circuit does not fall to zero when the current is zero. The value of B_i at *c* is referred to as the **remanence** of the iron. In order to reduce the flux in the circuit to zero it is necessary to reverse the current in the coil. The value of H_i needed to make B_i zero (the point *d*) is known as the **coercive force.** Increasing the current in the coil beyond the level needed to reach *d* eventually produces saturation of the iron at *e* with the direction of the flux opposite to that at *b*. Finally, reducing the current to zero and then increasing it with reversed polarity produces the curve *efgb*. If the current in the coil is repeatedly taken through the same cycle, the **hysteresis loop** *bcdefgb* is traversed repeatedly in a stable manner, provided only that the maximum value of the current in the coil is the same for both polarities. In many applications the flux in the iron is produced by a periodic (though not necessarily sinusoidal) current, and the iron then behaves in the manner described. If the maximum current is increased still further, there is no change to the loop because the magnetization of the iron cannot be increased beyond its value at saturation. If, on the other hand, a periodic current of smaller amplitude is used, it is found that the behaviour of the iron is described by a smaller loop such as the one shown dotted in Fig. 5.12.

The properties of magnetic materials can be varied by making alloys with different proportions. They are also strongly influenced by the ways in which the finished material is prepared, especially any heat treatments used. A very great variety of materials now exists. Their properties can be looked up in reference books. From a practical point of view the main division among magnetic materials is between 'hard' and 'soft' materials. The differences between these two classes are illustrated in Fig. 5.13. Soft magnetic materials are characterized by having narrow hysteresis loops, low remanence and small coercive forces. They are therefore easily magnetized and demagnetized and are used as conductors of flux in magnetic circuits and magnetic screens. Because these materials have narrow hysteresis loops their behaviour can be approximated by their initial magnetization curves. Moreover, if the loop is also nearly straight for fields below saturation,

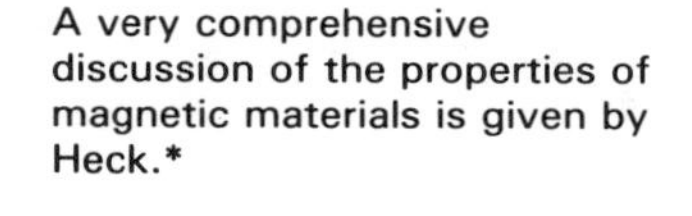
A very comprehensive discussion of the properties of magnetic materials is given by Heck.*

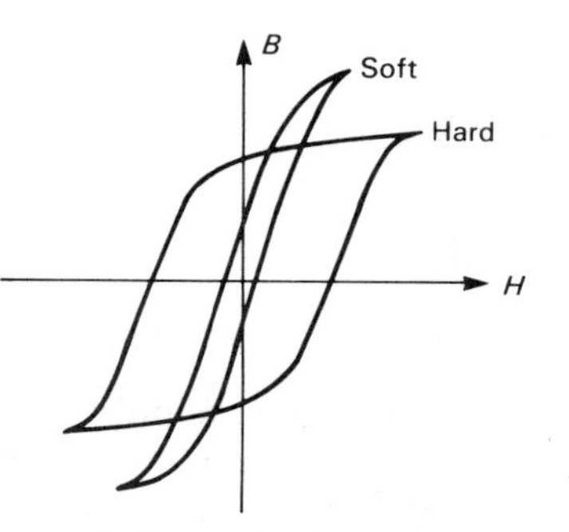

Fig. 5.13 Typical hysteresis loops for hard and soft magnetic materials.

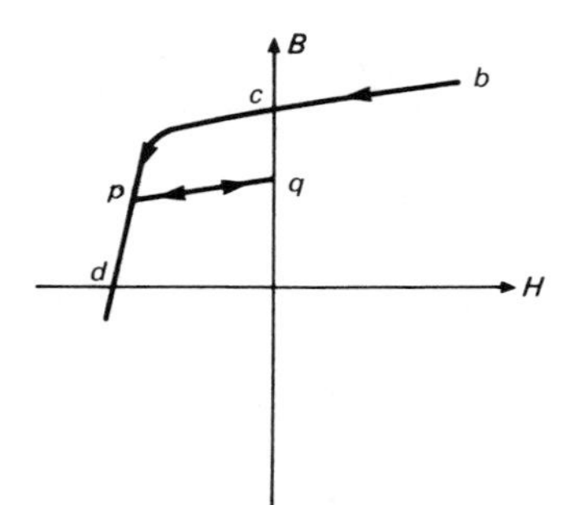

Fig. 5.14 Part of a B–H curve showing a minor loop p–q.

like the one shown in Fig. 5.12, it is possible to make the approximation that the permeability is constant.

Hard magnetic materials have broad hysteresis loops, high remanence and high coercive forces. Such materials are difficult to demagnetize. They are used for making permanent magnets and magnetic recording materials.

A further important feature of the properties of magnetic materials is illustrated by Fig. 5.14. Suppose that the material has been magnetized to saturation (b) and then demagnetized to the point p by the application of a reverse field. If the magnetizing field is then reduced to zero, the working point of the material moves to q along a **minor loop.** Minor loops are usually narrow and it is possible to approximate the loop between p and q by the straight line pq, which is usually nearly parallel to bc. Provided that the demagnetizing field applied does not take the working point past p back onto the main loop, the material will operate in a stable manner along the minor loop pq. This fact is of considerable importance in the application of permanent magnets.

Solution of problems in which μ cannot be regarded as constant

Consider again the magnetic circuit shown in Fig. 5.7. Neglecting leakage and fringing fluxes we may assume that $B_i = B_a$. The application of the magnetic circuit law to the circuit gives

$$\begin{aligned} NI &= 4LH_i + gH_a \\ &= 4LH_i + gB_i/\mu_0 \end{aligned}$$

or

$$B_i = (\mu_0/g)(NI - 4LH_i) \tag{5.18}$$

A second condition is that B_i and H_i must be related to each other by the hysteresis loop of the material:

$$B_i = f(H_i) \tag{5.19}$$

Equations (5.18) and (5.19) are a pair of simultaneous equations in B_i and H_i. However, because (5.19) is not the equation of a straight line, they are non-linear simultaneous equations. They can be solved graphically or by using an analytical approximation to the equation of the hysteresis loop. The graphical solution is

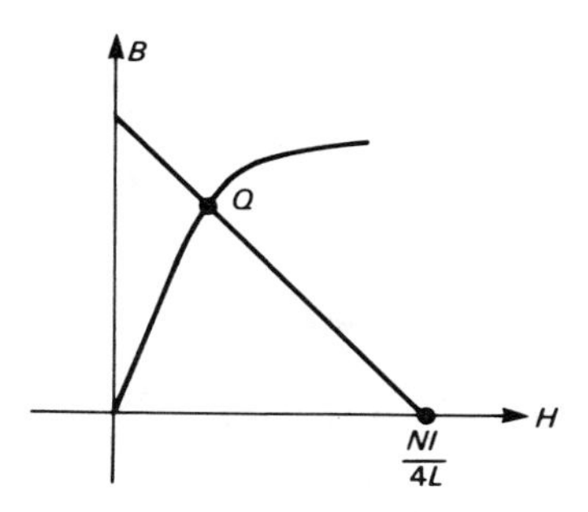

Fig. 5.15 A magnetic circuit problem can be solved by drawing a load line across the B–H plot for the iron. The working point of the circuit is at q.

shown in Fig. 5.15. The working point q is at the intersection of the straight line represented by Equation (5.18) with the hysteresis loop. The intercept of this line on the horizontal axis is the total magnetomotive force in the circuit, while its slope is determined by the relative sizes of the iron path and the air gap. The method is similar to the process of finding the working point of a transistor by drawing a load line across the characteristic curves.

The technique of finding the working point of a transistor by drawing a load line is discussed by Ritchie, G.J., *Transistor Circuit Techniques*, 3rd edn, Chapman & Hall (1992), p. 18.

Worked Example 5.3

A magnetic circuit like that shown in Fig. 5.7 has the following parameters

$L = 50$ mm $\quad g = 0.2$ mm
$w = 10$ mm
$N = 100$ turns $\quad I = 1.0$ A

Given that the core is made of a magnetic material having the initial magnetization curve shown in Fig. 5.16, estimate the flux density in the air gap.

Solution We assume that for this material the initial magnetization curve is an adequate approximation to the hysteresis loop. Allowing for the fringing around the air gap, $\Phi = (w + g)^2 B_a = w^2 B_i$, so that

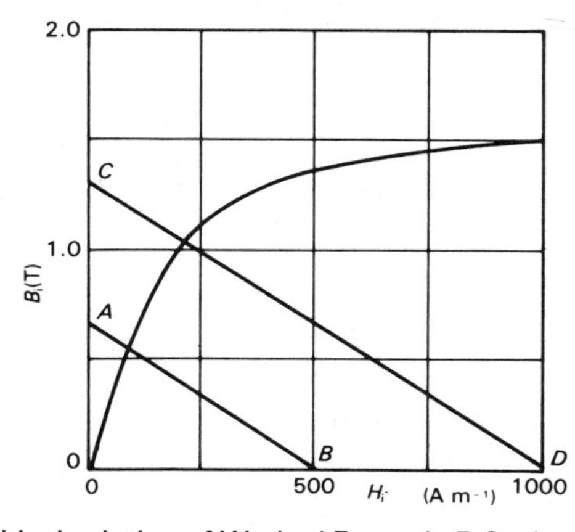

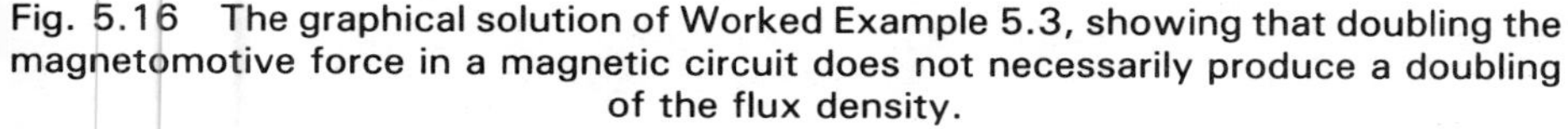
Fig. 5.16 The graphical solution of Worked Example 5.3, showing that doubling the magnetomotive force in a magnetic circuit does not necessarily produce a doubling of the flux density.

$$H_i = (NI - gB_a/\mu_0)/4L$$
$$= (100 - 153B_i)/0.2$$
$$= (500 - 765B_i)$$

or

$$B_i = (500 - H_i)/765$$

This load line is plotted as AB on Fig. 5.16. The working point is

$$H_i = 85 \text{ A m}^{-1}, \qquad B_i = 0.55 \text{ T}$$

and the flux density in the air gap is

$$B_a = 0.53 \text{ T}$$

B_a is less than B_i because of the fringing field of the gap.

If the current in the coil is doubled so that the load line is CD, then the working point is

$$H_i = 210 \text{ A m}^{-1} \qquad B_i = 1.04 \text{ T}$$

and

$$B_a = 1.0 \text{ T}$$

Doubling the current in the coil does not result in a doubling of the flux density in the air gap. This shows clearly the non-linear behaviour of the material.

When a magnetic circuit is being designed it is likely that the reverse problem has to be solved. The flux is specified and the ampere turns to produce it have to be calculated.

Worked Example 5.4 Calculate the ampere turns needed to produce a flux of 0.14×10^{-3} Wb in the magnetic circuit of Worked Example 5.3.

Solution The cross-sectional area of the iron is 10^{-4} m^2 so the flux density in the iron is $B_i = 1.4$ T. The magnetic field strength in the iron is read from Fig. 5.16: $H_i = 510$ A m^{-1}. Neglecting fringing, $B_a = B_i$ and $H_a = B_a/\mu_0 = 1.1 \times 10^6$ A m^{-1}. Applying the magnetic circuit law, the total number of ampere turns is

$$NI = H_i l_i + H_a l_a$$
$$= 510 \times 0.2 + 1.1 \times 10^6 \times 0.2 \times 10^{-3}$$
$$= 322$$

The number of turns used would depend upon the current available.

Permanent magnets

We have already noted that some magnetic materials can be magnetized so that they produce a substantial magnetic flux even when the magnetizing field is removed. These 'hard' magnetic materials are used for making **permanent magnets**. Permanent magnets are used in loudspeakers and in a variety of microwave devices such as ferrite isolators, magnetron oscillators and travelling-wave tube amplifiers. Figure 5.17 shows a typical B–H plot for a permanent magnet material. This curve, which is the part of the hysteresis loop lying in the second quadrant, is known as the **demagnetization curve** of the material. Suppose that the circuit shown in Fig. 5.7 is made of this material and magnetized to saturation by passing a current through the coil with the air gap bridged by a piece of soft iron. When the current in the coil is reduced to zero the working point will be at P in Fig. 5.17. Neglecting fringing and leakage fluxes we have, when the soft iron is removed from the gap,

Ferrites are manufactured from oxides of iron. They can be made as both hard and soft magnetic materials. Magnetically soft ferrites are used for magnetic cores at high frequencies, for reasons which are discussed in Chapter 6.

$$B_i = -(4L/g)\mu_0 H_i \tag{5.20}$$

from Equation (5.18), setting $I = 0$. This load line passes through the origin and has negative slope. It is plotted in Fig. 5.17 as the line OQ. The effect of opening the gap is therefore to demagnetize the magnet to some extent. For this reason it is common to speak of the **demagnetizing field** of the air gap.

If the piece of soft iron is reinserted in the air gap the operating point of the magnet moves along a minor loop to the point P'. In this state the magnet is largely immune to the effects of external fields. For this reason it is usual to store permanent magnets with their air gaps bridged by a piece of soft iron known as a 'keeper'.

Suppose that someone has forgotten to put a keeper on the magnet and that it is exposed to a magnetic field which tends to demagnetize it. The working point might then move to R. When the field is removed the working point lies at Q' on the minor loop RS. This illustration shows that it is necessary to treat permanent magnets carefully if their properties are not to be affected. For example, it is unwise to try to force a pair of magnets together with their fields opposing because they are then trying to demagnetize each other. It is not a good thing that the working point of the magnet should be sensitive to external influences in this way. For

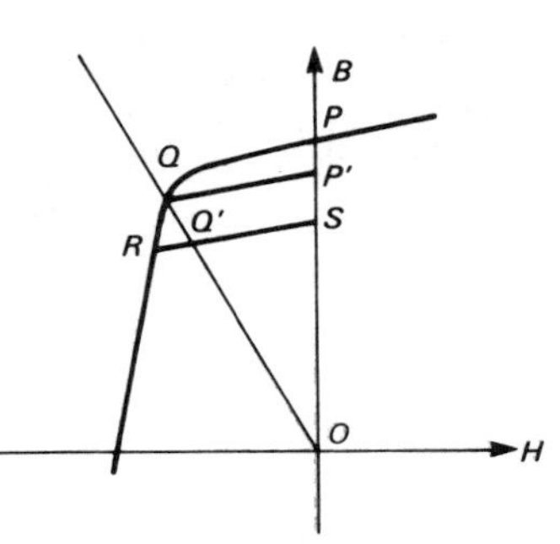

Fig. 5.17 The working point for a permanent magnet circuit lies in the second quadrant of the B–H plot at a point such as Q. A permanent magnet can be stabilized against demagnetization by external fields by operating it at a point Q' on a minor loop such as R–S.

this reason permanent magnets are often stabilized by deliberately demagnetizing them beyond their working points on the main hysteresis loop. The operation of magnets which have been stabilized is stable along a recoil line such as RS unless the external demagnetizing field is very strong. Smaller external fields produce a temporary shift in the working point, but it returns to Q' when the perturbing field is removed.

Worked Example 5.5 Figure 5.18(b) shows a cross-sectional view of a cylindrical loudspeaker magnet. Estimate the flux density in the air gap.

Solution To get a rough estimate of the flux density in the air gap we neglect fringing, neglect the reluctance of the soft iron parts of the circuit, and assume that the magnet is operating on its main demagnetization curve. From the magnetic circuit law we know that

$$H_i = -(l_a/l_i)H_a = -(1/20)H_a = -H_a/20$$

The condition for the continuity of flux gives

$$B_i = (A_a/A_i)B_a = ((20 \times 3\pi)/\pi(25^2 - 15^2))$$
$$= 3\,B_a/20$$

But $B_a = \mu_0 H_a$, so that $B_i = -3\mu_0 H_i = -3.8 \times 10^{-6} H_i$.

Figure 5.19 shows the demagnetization curve for Feroba 1 with the load line plotted on it. The working point of the magnet is $H = 40 \times 10^3$ A m^{-1}, $B = 0.15$ T. The flux density in the air gap is 1.0 T.

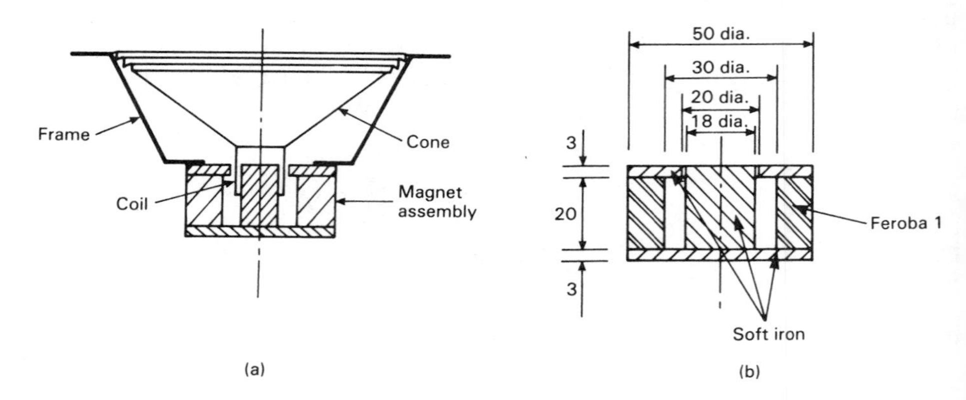

Fig. 5.18 (a) A cross-sectional view of a moving-coil loudspeaker. The coil is placed in the air gap of the magnetic circuit so that it experiences an axial force when a current is passed through it. The movement of the coil is transferred to the air through the motion of the paper cone. This is one of the very few examples of the use of Equation (4.19) in electronic engineering. (b) An enlarged view of the magnet assembly. For an alternative magnet arrangement see Fig. 5.26.

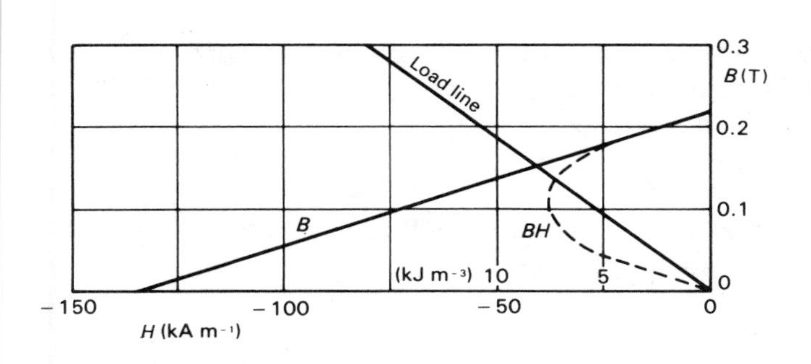

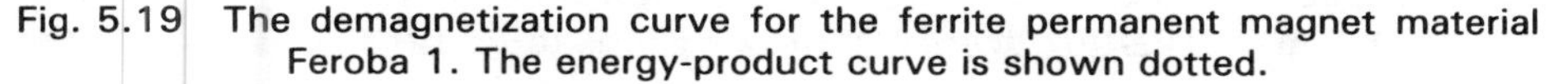
Fig. 5.19 The demagnetization curve for the ferrite permanent magnet material Feroba 1. The energy-product curve is shown dotted.

Using permanent magnets efficiently

In a circuit involving a permanent magnet, the cost of the magnet is usually a considerable part of the cost of the whole circuit. It is, therefore, desirable to use the magnet material as efficiently as possible. A rule of thumb for this purpose can be derived by considering the circuit shown in Fig. 5.7. Let the lengths of the magnet and the air gap be l_i and l_a and let their cross-sectional areas be A_i and A_a. The magnetic circuit law gives

$$l_i H_i + l_a H_a = 0 \tag{5.21}$$

and the conservation of flux gives

$$A_i B_i = A_a B_a \tag{5.22}$$

The volume of the magnet is

$$\begin{aligned} V_i &= l_i A_i \\ &= -l_a A_a \, (H_a B_a / H_i B_i) \end{aligned}$$

Now $l_a A_a$ is the volume of the air gap which may be taken as given, and H_a and B_a are the field strength and flux density in the air gap. So, the minimum magnet volume to provide a given flux density in a given air gap is achieved by making the product $H_i B_i$ as great as possible. This is known as **Evershed's criterion**. The quantity $H_i B_i$ is known as the **energy product** of the magnet for reasons which will become apparent in a later chapter. Figure 5.20 shows the curves of B against

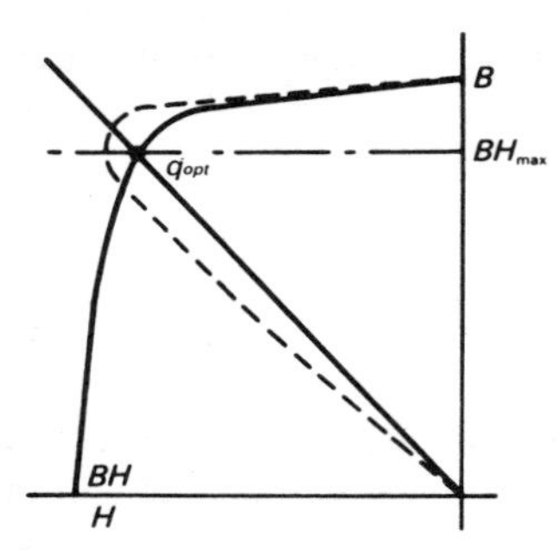

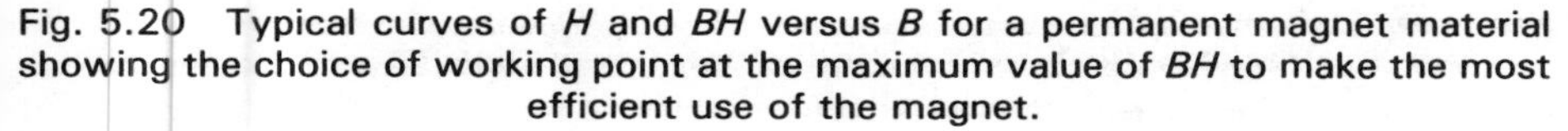
Fig. 5.20 Typical curves of H and BH versus B for a permanent magnet material showing the choice of working point at the maximum value of BH to make the most efficient use of the magnet.

H and BH against B as they are usually plotted. The optimum working point of the material is then that which gives the value of B corresponding to the maximum value of BH as shown at Q_{opt}.

Worked Example 5.6 Redesign the magnetic circuit shown in Fig. 5.18(b) so that the magnetic material is used as efficiently as possible.

Solution Figure 5.19 shows the demagnetization curve of Feroba 1 with the curve of BH against B added. The energy product is a maximum when $B = 0.11$ T and $H = 68$ kA m^{-1}. The flux density in the air gap is to be 1.0 T as before, so the cross-sectional area of the magnet must be 9 times that of the air gap. Thus $A_i = 9 \times 60\pi$ mm^2. Assuming that the inner diameter of the magnet is to remain unchanged, it can be shown that the outer diameter must be 55 mm, to give the correct area. Similarly the length of the magnet must be $(H_a/H_i)l_a = (800/68) \times 1.0 = 12$ mm.

Summary

In this chapter we have considered the properties of ferromagnetic materials. We have seen that their principal characteristic is that they acquire a very strong magnetization when placed in a magnetic field. As with dielectric materials it is possible to represent the properties of the materials by introducing a new vector, in this case $\boldsymbol{H}$, which includes the effects of the polarization of the materials. Using the vector $\boldsymbol{H}$ a new form of the magnetic circuit law was derived, which is valid even when magnetic materials are present. Consideration of the boundary conditions at the surface of a ferromagnetic material showed that such materials can usually be regarded as good conductors of magnetic flux. This property allows the path of the flux to be controlled so that it is directed to the place where it is needed, as in a recording head, or directed away from sensitive areas in magnetic screening. In a limited range of cases it is possible to regard ferromagnetic materials as being linear, but, in general, it is necessary to take account of their non-linear behaviour as described by their hysteresis loops. The method of making calculations for simple magnetic circuits has been illustrated in a number of worked examples covering both electromagnets and permanent magnets.

Problems

Solenoids very like this are used for collimating the high-power electron beams in klystrons for television transmitters and other microwave tubes (see Problem 4.9).

5.1 A solenoid of the form shown in Fig. 5.21 may be idealized as a set of circular current loops at the centres of the windings. Estimate the maximum and minimum values of the magnetic flux density on the axis of the solenoid. What is the effect of changing the coil spacing to $2a$?

5.2 Figure 5.22 shows a 'pot core' made of Ferroxcube B4. Cores like this are used to make radio-frequency inductors. Given that the coil has 100 turns, estimate the maximum a.c. current which can be passed through it without

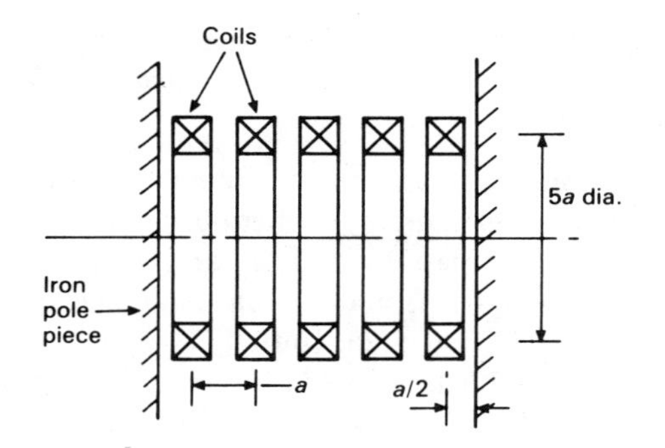

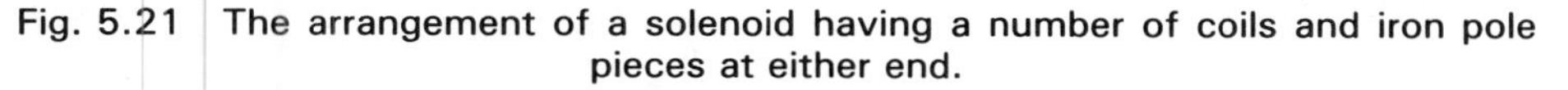
Fig. 5.21 The arrangement of a solenoid having a number of coils and iron pole pieces at either end.

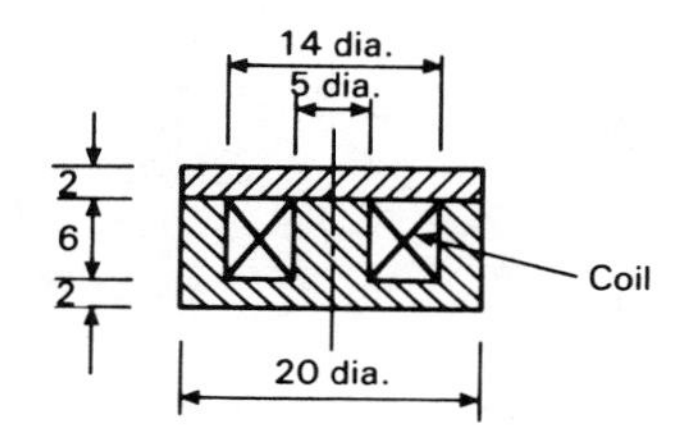

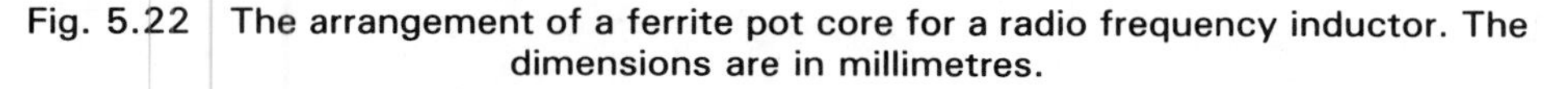
Fig. 5.22 The arrangement of a ferrite pot core for a radio frequency inductor. The dimensions are in millimetres.

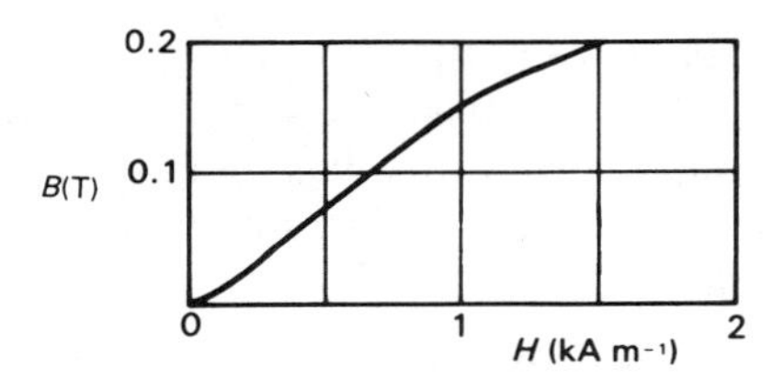

Fig. 5.23 The initial magnetization curve for the ferrite material Ferroxcube B4.

the behaviour of the inductor becoming non-linear. The initial magnetization curve of Ferroxcube B4 is shown in Fig. 5.23.

5.3 Figure 5.24 shows a magnetic recording head made of Permalloy D. Given

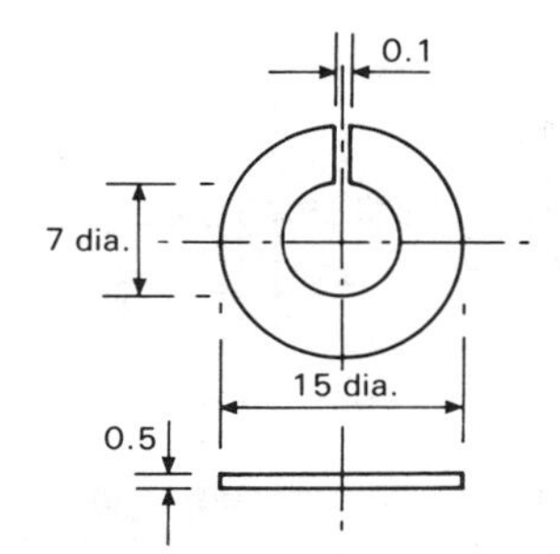

Fig. 5.24 The magnetic core for a recording head. The dimensions are in millimetres.

that the magnetic material has a relative permeability of 500, estimate the number of turns needed to produce a flux density of 0.05 T in the air gap from a current of 10 mA.

5.4 A magnetic circuit consists of a soft iron yoke, which may be assumed to have infinite permeability, a permanent magnet 60 mm long and 2400 mm^2 cross-sectional area, and an air gap 5 mm long and 3600 mm^2 in cross-sectional area. The permanent magnet is made from Columax, which has the demagnetization curve shown in Fig. 5.25. Initially a keeper made of soft magnetic material is inserted into the air gap and the magnet magnetized to a residual flux density of 1.35 T by means of a coil wound on the iron yoke.

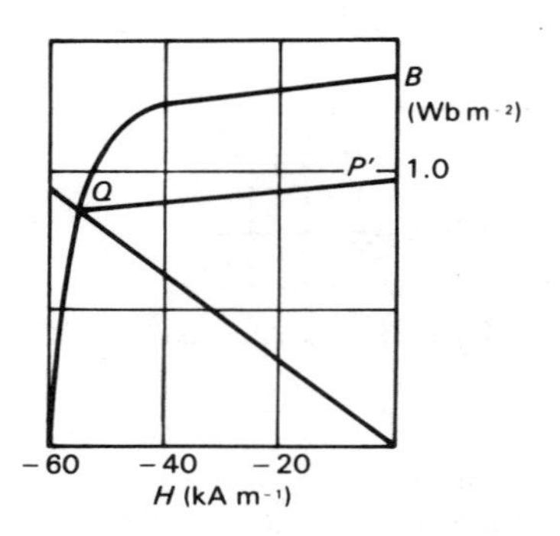

Fig. 5.25 The demagnetization curve for Columax.

(a) Determine the flux density in the air gap when the keeper is removed if fringing around the gap can be neglected.
(b) Determine the value of the flux density in the permanent magnet if the keeper is replaced in the air gap, assuming that the recoil permeability of Columax is $1.8\mu_0$.

5.5 Figure 5.26 shows a cross-sectional view of a cylindrical loudspeaker magnet. The permanent magnet is made from Columax (Fig. 5.25) and operates on its demagnetization curve. Neglecting fringing fields, leakage flux, and the reluctance of the soft iron pole pieces estimate the flux density in the air gap. Calculate the optimum magnet dimensions for this flux density. If the magnet were already in production would you consider this design change justified?

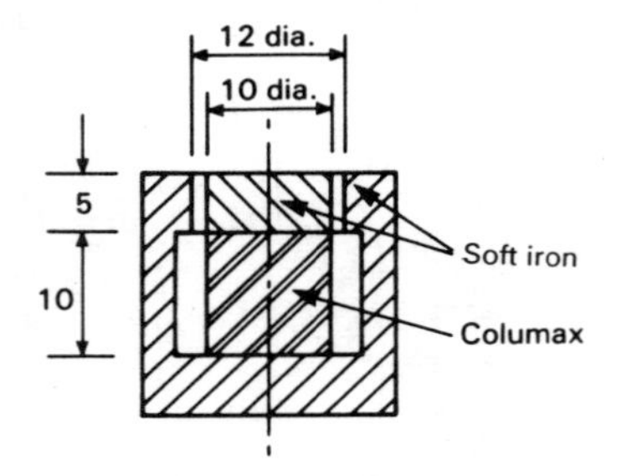

Fig. 5.26 The arrangement of a Columax loudspeaker magnet. Notice how the use of a magnet material with a higher-energy product allows the permanent magnet to be put at the centre instead of outside as shown in Fig. 5.18(b).

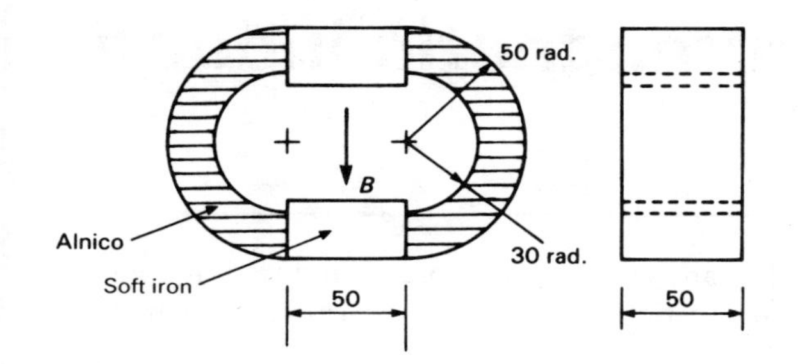

Fig. 5.27 The arrangement of an Alnico magnet for a magnetron. This magnet is typical of those used for these radar transmitter tubes.

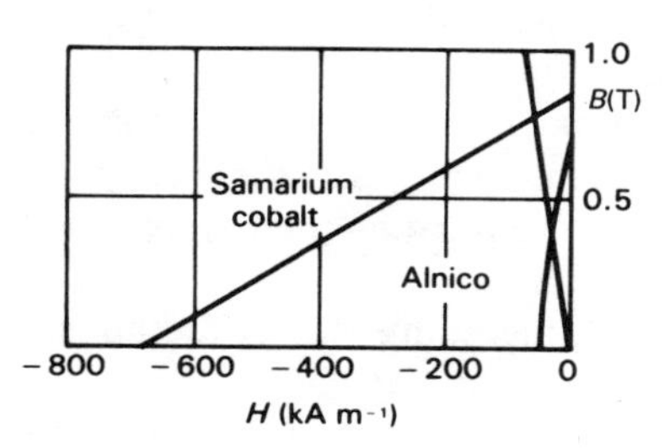

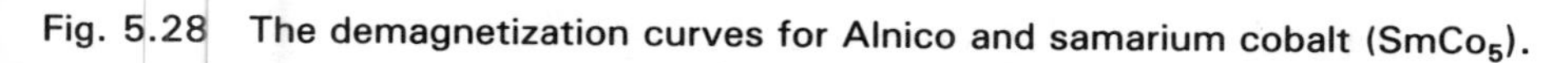

Fig. 5.28 The demagnetization curves for Alnico and samarium cobalt ($SmCo_5$).

5.6 Figure 5.27 shows an Alnico magnet for a magnetron, which is operated at a flux density of 0.4 T. How much weight could be saved by replacing it with a samarium cobalt magnet? The demagnetization curves of the two materials are shown in Fig. 5.28. The density of Alnico is 7300 kg m^{-3} and that of samarium cobalt is 8100 kg m^{-3}.

6 Electromagnetic induction

Objectives

- ☐ To show that an electromotive force is induced in a conductor moving through a magnetic field, and that a current flows if the circuit is completed.
- ☐ To show that an electromotive force is induced in a loop of wire moving through a non-uniform magnetic field and that, by implication, an e.m.f. is induced in a loop of wire when the magnetic flux linked to it changes.
- ☐ To generalize the first two objectives in the form of Faraday's law of electromagnetic induction.
- ☐ To demonstrate the links between the circuit concepts of self- and mutual inductance and electromagnetic field theory.
- ☐ To discuss the causes of electromagnetic interference and ways of reducing it.
- ☐ To introduce methods, including energy methods, for calculating self- and mutual inductances.
- ☐ To introduce the analogy betwen L, C and R in two-dimensional problems.
- ☐ To discuss the concept of the storage of energy in a magnetic field, and to show the equivalence between calculations of the energy stored in an inductor from field and circuit points of view.
- ☐ To discuss the special case of energy storage in a magnetic field in iron anc to introduce the idea of hysteresis loss.
- ☐ To introduce the idea of the induction of eddy currents in a conductor by changing magnetic field.

In this chapter we turn our attention to phenomena involving conductors an magnetic fields in relative motion, and magnetic fields which are changing wit time. These effects, first investigated by Faraday in 1831, complete the lin between magnetism and electricity. We have already seen how magnetic fields ar produced by electric currents. We shall now show that electromotive forces an electric currents can be produced by electromagnetic induction. These ideas ar fundamental to the generation of electricity by electromagnetic machines and t the use of transformers in the distribution of electric power. As far as electroni engineers are concerned their importance lies in the fact that they provide a explanation of self- and mutual inductance, and the means to calculate thes circuit parameters.

The current induced in a conductor moving through a steady magnetic field

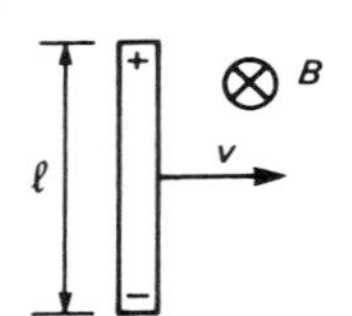

Consider a piece of straight wire moving with constant velocity v through a stead magnetic field B which is at right angles to the wire and to its direction of motion as shown in the figure in the margin. The conduction electrons in the wire ar carried through the field with the wire so, from Equation (4.4), each electro experiences a magnetic force qvB acting upon it. The cross product symbols have been omitted because the directions of the wire, the velocity of the wire and the field were chosen to be mutually perpendicular. As a result of this force the electrons are displaced so that the ends of the wire acquire charges, as shown in the figure in the margin. The polarization of the wire produces an electric field within it which opposes the motion of the electrons. In equilibrium the electric force on each charge must be equal and opposite to the magnetic force, so that

$$E = vB \tag{6.1}$$

The potential difference between the ends of the wire is therefore

$$V = \int E\,\mathrm{d}l = vBl \tag{6.2}$$

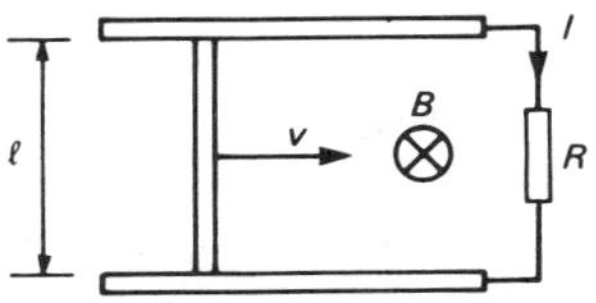

Fig. 6.1 A current is induced in a closed circuit in a magnetic field if the shape of the circuit is changed.

Now suppose that the ends of the wire are connected to parallel rails by sliding contacts and that the circuit is completed by a load resistor, as shown in Fig. 6.1. The potential difference between the ends of the rod acts as an electromotive force in the circuit and a steady current flows such that

$$IR = vBl \tag{6.3}$$

The electric power dissipated in the resistor is

$$I^2R = vblI \tag{6.4}$$

The mechanical power input is the product of the force needed to move the wire and its velocity. Since the current in the circuit is I, the force needed to move the wire is, from equation (4.19), IlB, so the mechanical power input is

$$Fv = vBlI \tag{6.5}$$

This is exactly equal to the electric power, demonstrating that this device converts mechanical energy into electrical energy. Examination of Fig. 6.1 shows that the force exerted on the current-carrying slider by the field is in the opposite direction to v. This internal force must be opposed by an equal and opposite external force if the wire is in steady motion. The external force is in the same direction as v, so it does work on the slider, providing a power input to the device. This simple arrangement forms the starting point for the discussion of the generation of electricity by electromagnetic machines. Important as the topic is, it is outside the scope of this book and will not be pursued any further.

For further information about the application of electromagnetic theory to electric machines, consult Slemon.*

Recalling the definition of magnetic flux which was used in earlier chapters, we can see that, with the direction of velocity shown, the flux linked to the circuit of Fig. 6.1 is changing at a rate given by

$$\frac{d\Phi}{dt} = -vlB \tag{6.6}$$

but, from Equation (6.2), this is just the electromotive force developed in the circuit. The direction of the electromotive force is such that when a current flows in the circuit the magnetic forces oppose the external mechanical force which is causing the flux to change. These ideas can be combined by writing

$$\mathcal{E} = -\frac{d\Phi}{dt} \tag{6.7}$$

The current induced in a loop of wire moving through a non-uniform magnetic field

In the experiment described in the previous section it was shown that a current could be induced in a circuit by changing the flux linked to the circuit by altering the shape of the circuit. It is interesting to see whether a current could also be induced by keeping the shape of the circuit fixed and varying the strength of the magnetic field. This could be achieved by taking a square loop of wire, as shown in Fig. 6.2(a), and moving it with a steady velocity v through a magnetic field perpendicular to the plane of the loop whose variation with position in the direction of motion is given by the graph of Fig. 6.2(b). At the instant when the wire AD is at x relative to the origin, the e.m.f. generated in it is

$$\mathcal{E}_{AD} = vlB_0(x/l) \tag{6.8}$$

where the order of the subscripts shows that the polarity of the e.m.f. is such as to drive a current from A to D through the rest of the circuit. In the same way the e.m.f. generated in BC is

$$\mathcal{E}_{BC} = vlB_0(x + l)/l \tag{6.9}$$

The remaining sides of the loop are moving parallel to their own directions so no potential differences are generated between their ends. The net e.m.f. in the circuit is therefore

$$\begin{aligned}\mathcal{E} &= vlB_0(x + l)/l - vlB_0x/l \\ &= vlB_0\end{aligned} \tag{6.10}$$

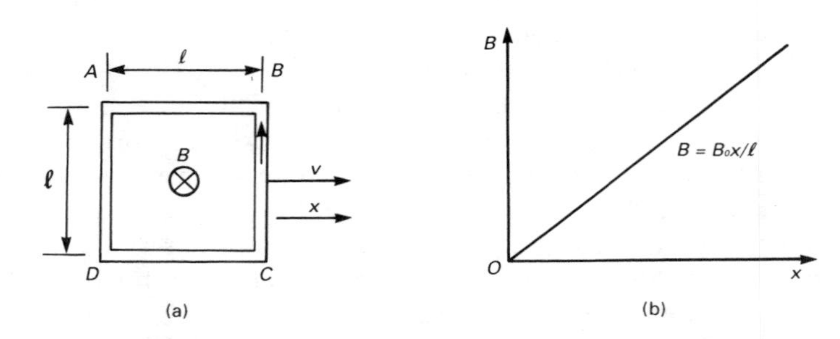

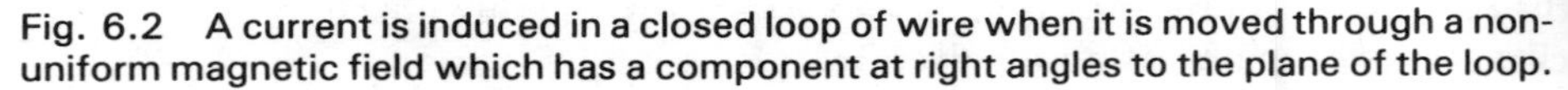
Fig. 6.2 A current is induced in a closed loop of wire when it is moved through a non-uniform magnetic field which has a component at right angles to the plane of the loop.

directed so as to tend to drive a current in a counterclockwise direction. Now the flux linked to the circuit is

$$\Phi = l^2 B_0 (x + \tfrac{1}{2} l)/l \qquad (6.11)$$

so that

$$\frac{\mathrm{d}\Phi}{\mathrm{d}t} = l B_0 \frac{\mathrm{d}x}{\mathrm{d}t} = v l B_0 \qquad (6.12)$$

So, in this case also, the e.m.f. is equal to the rate of change of flux linkage. The direction of the flux produced by the induced current is opposed to the original field, so that it tends to prevent the flux linked to the loop from increasing as the loop is moved. Since it is the relative motion of the loop and the field which produces the e.m.f. it is reasonable to suppose that the same result would have been produced if the loop had been held fixed and the source of the field moved with velocity v in the negative x-direction.

Faraday's law of electromagnetic induction

In the previous sections it has been shown that, at least in the two cases considered, an electromotive force is induced in a circuit when the magnetic flux linked to the circuit is changed. This phenomenon was observed experimentally by Faraday, who generalized his findings in the law which bears his name. This law says that, *if the flux linked to a circuit is changed in any way, then an electromotive force is induced in the circuit whose magnitude is proportional to the rate of change of the flux linkage to the circuit*. The term **flux linkage**, is used here to include circuits such as coils in which the wire encircles the path of the circuit more than once. In these cases the flux linked to the circuit is the sum of the flux linkage to the individual turns. It is often possible to make the approximation that the flux linkage to each turn of a coil is the same, so that the total flux linkage is the product of the flux linked to one turn and the number of turns. Denoting flux linkage by Λ, this relationship is expressed by the equation

The definition of 'flux linkage' is not always a straightforward matter. A useful discussion of difficulties in interpreting the laws of electromagnetic induction is given by G.W. Carter.*

$$\Lambda = N\Phi \qquad (6.13)$$

We have also seen that the direction of the e.m.f. induced in a circuit by a changing magnetic flux linkage is always such that it tries to oppose the change of flux which causes it. This statement is known as **Lenz's law**.

Faraday's law can be expressed in mathematical terms by writing

$$\mathcal{E} = -\frac{\mathrm{d}\Lambda}{\mathrm{d}t} \qquad (6.14)$$

For a simple circuit such as that shown in Fig. 6.3, the flux linkage is given by

$$\Lambda = \int_S \boldsymbol{B} \cdot \mathbf{d}\boldsymbol{A} \qquad (6.15)$$

where the integration is carried out over an open surface spanning the circuit and the elementary vectors $\mathbf{d}\boldsymbol{A}$ are taken to point upwards as shown. The e.m.f. in the circuit can be calculated in a similar manner by making use of Equation (3.17).

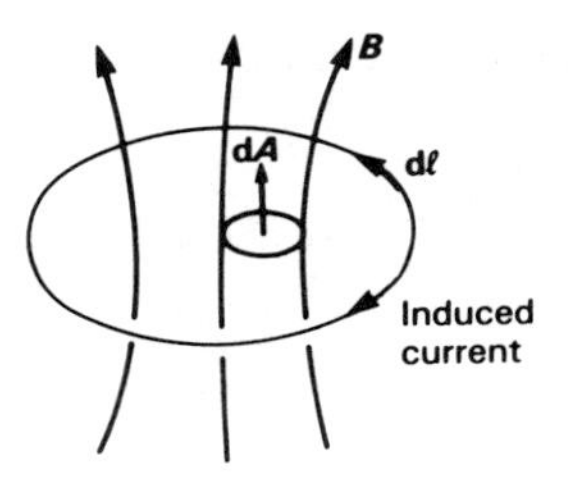

Fig. 6.3 The direction of the current induced in a loop of wire by a magnetic flux which is increasing with time.

The direction of the line integral around the circuit is chosen to be in a right-hand corkscrew sense with respect to the direction of the vector $\mathbf{d}\boldsymbol{A}$. This direction is shown in Fig. 6.3 by the direction of the elementary vector $\mathbf{d}\boldsymbol{l}$. Thus

$$\mathcal{E} = \oint \boldsymbol{E}\cdot\mathbf{d}\boldsymbol{l} \tag{6.16}$$

Now, if $\boldsymbol{B}$ is in the direction shown and it is increasing in magnitude with time, Lenz's law requires the direction of the induced current to be opposite to the direction of the line integral as shown in Fig. 6.3. Substituting for $\mathcal{E}$ and Λ in Equation (6.14) produces a general mathematical form of Faraday's law:

$$\oint \boldsymbol{E}\cdot\mathbf{d}\boldsymbol{l} = -\frac{\mathrm{d}}{\mathrm{d}t}\iint \boldsymbol{B}\cdot\mathbf{d}\boldsymbol{A} \tag{6.17}$$

Although this expression may seem abstract and complicated it should be remembered that it is no more than a restatement of Faraday's law and Lenz's law in the symbolic language of mathematics. When a circuit has more than one turn the integral on the right-hand side of Equation (6.17) is taken over all the turns. In the problems which are of interest in electronics the circuit is normally fixed and the magnetic field changing with time. Under these circumstances the differentiation can be brought inside the integral to give

$$\oint \boldsymbol{E}\cdot\mathbf{d}\boldsymbol{l} = -\iint \frac{\partial \boldsymbol{B}}{\partial t}\cdot\mathbf{d}\boldsymbol{A} \tag{6.18}$$

Inductance

We are now in a position to consider the third of the trio of passive circuit parameters, resistance, capacitance and inductance. Figure 6.4 shows two loops

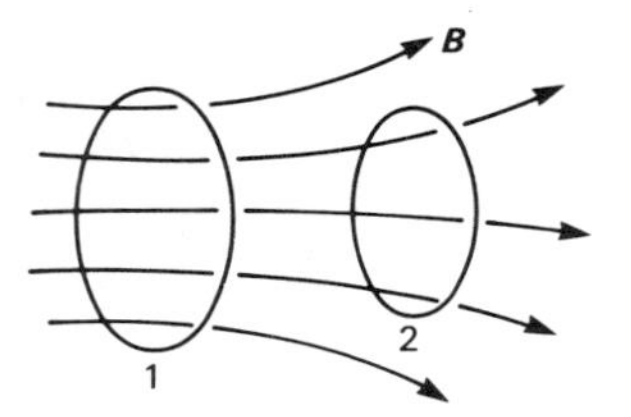

Fig. 6.4 Two loops of wire are inductively linked when some or all of the magnetic flux produced by one of them passes through the other.

of wire arranged so that when a current is passed through loop 1 some of the flux produced passes through loop 2 and vice versa. Suppose that the flux linked to loop 2 is Λ_{21} when the current in loop 1 is I_1. If the loops are in a region of constant permeability, then Λ_{21} is proportional to I_1. The constant of proportionality is

$$L_{21} = \Lambda_{21}/I_1 \tag{6.19}$$

L_{21} is the **mutual inductance** of the loops. From Equation (6.19) it is evident that inductance has the dimensions Webers per ampere, and this unit is called the **henry**. The ratio of the flux linked to loop 1 to the current in the same loop is the **self-inductance** of that loop, given by

$$L_{11} = \Lambda_1/I_1 \tag{6.20}$$

Two similar equations can be derived by assuming that the source current is in loop 2. These equations can then be brought together by superposition to give

$$\begin{aligned} \Lambda_l &= \Lambda_{11} + \Lambda_{12} = L_{11}I_1 + L_{12}I_2 \\ \Lambda_2 &= \Lambda_{21} + \Lambda_{22} = L_{21}I_1 + L_{22}I_2 \end{aligned} \tag{6.21}$$

This is a passive linear system, so the matrix of the coefficients of inductance must be symmetrical; that is

$$L_{21} = L_{12} \tag{6.22}$$

The proof that $L_{12} = L_{21}$ is given by Bleaney and Bleaney.*

This is the mutual inductance of the loops often represented by the symbol M. Using this notation the double subscripts can be dropped and the self-inductances represented by the symbols L_1 and L_2. When Faraday's law is applied to equations (6.21), we obtain

$$\begin{aligned} \mathcal{E}_1 &= -L_1\frac{\mathrm{d}I_1}{\mathrm{d}t} - M\frac{\mathrm{d}I_2}{\mathrm{d}t} \\ \mathcal{E}_2 &= -M\frac{\mathrm{d}I_1}{\mathrm{d}t} - L_2\frac{\mathrm{d}I_2}{\mathrm{d}t} \end{aligned} \tag{6.23}$$

where the positive directions of the currents and the e.m.f.s are the same. These are the familiar circuit equations for inductively coupled circuits. It is important to remember that these equations hold only when the system is linear, that is, the fluxes are directly proportional to the currents producing them. Since the behaviour of iron can be highly non-linear, as we saw in Chapter 5, it cannot be taken for granted that iron-cored inductors behave as linear circuit elements.

The self- and mutual inductances in a circuit are related to each other, as will now be demonstrated. Consider again the pair of loops shown in Fig. 6.4, but suppose that loop 1 is a coil of N_1 turns and coil 2 has N_2 turns. When a current I_1 flows in 1 a flux Φ_1 is generated. For the sake of simplicity it will be assumed that the whole of this flux is linked to every turn in coil 1, so that

$$\Lambda_1 = N_1\Phi_1 \tag{6.24}$$

In general, some of the flux will not pass through coil 2 so that the flux linked to that coil can be written

$$\Lambda_2 = k_2N_2\Phi_1 \tag{6.25}$$

where k_2 is the proportion of the flux generated by coil 1 which is linked to coil 2. It follows that $k_2 \leqslant 1$. Assuming linearity, we can write

$$M/L_1 = \Lambda_2/\Lambda_1 = k_2 N_2/N_1 \tag{6.26}$$

If a current is passed through coil 2 we obtain, by a similar argument

$$M/L_2 = \Lambda_1/\Lambda_2 = k_1 N_1/N_2 \tag{6.27}$$

where k_1 is the fraction of the flux produced by coil 2 which passes through coil 1. Multiplying Equations (6.26) and (6.27) together gives

$$M^2 = k_1 k_2 L_1 L_2$$

or $$M = k\surd(L_1 L_2) \tag{6.28}$$

where the coupling coefficient $k = \surd(k_1 k_2) \leqslant 1$

Electromagnetic interference

For more information about techniques for dealing with interference, consult Morrison.*

In Chapter 2 we saw that there can be unwanted capacitive coupling between two circuits. Similar problems can arise through inductive coupling. Figure 6.5 shows a simple example of the coupling of two circuits through their mutual inductance M. If an alternating current flows in circuit 1, then it produces a magnetic flux, some of which may be linked to circuit 2. To simplify matters we will assume that I_1 is much greater than I_2 so that, approximately,

$$I_1 = V_1 \exp(\mathrm{j}\omega_1 t)/(R_{s1} + R_{L1})$$

and the e.m.f. induced in circuit 2 is

$$\begin{aligned} V_i &= -M\frac{\mathrm{d}I_1}{\mathrm{d}t} \\ &= -\frac{\mathrm{j}\omega_1 M V_1}{(R_{s1} + R_{L1})}\exp(\mathrm{j}\omega_1 t) \end{aligned} \tag{6.29}$$

This voltage is added to the signal voltage in circuit 2. The interference is worst when V_1 is large and the loop impedance of circuit 1 is small. The commonest example of this is when circuit 1 represents the a.c. mains and the interference appears as 50 Hz 'mains hum'. In this case it is not possible to reduce the interference by altering V_1 or $(R_{s1} + R_{L1})$, so effort has to be concentrated on

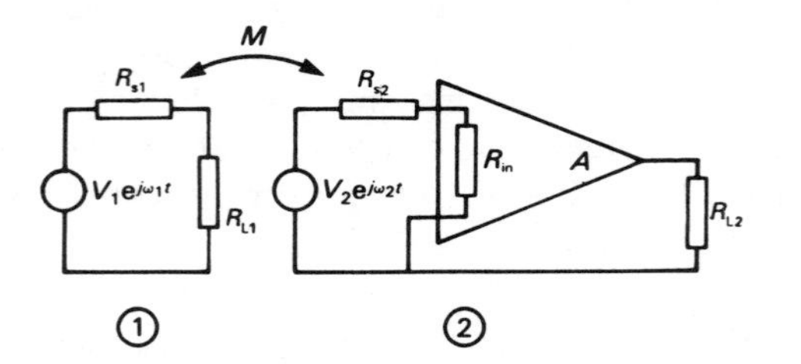

Fig. 6.5 When two circuits are coupled by stray mutual inductance the signal in one circuit can interfere with that in the other. For example, if circuit 1 is the a.c. mains a spurious 50 Hz signal could be added to the signal V_2 in circuit 2.

reducing M. To do this we have to reduce the proportion of the flux produced by circuit 1 which is linked into circuit 2. Three strategies are possible:

This subject is also discussed by Sangwine.*

1. *Reduce the area of circuit 2.* If the signal source V_2 is a transducer some distance from the amplifier, then this can be achieved by using a pair of wires twisted together (a 'twisted pair') to connect them together.
2. *Rotate circuit 2 so that its plane is parallel to the flux produced by circuit 1.* Rotating one circuit relative to the other can reduce the flux linkage to zero. If the two circuits are to be enclosed in the same case in fixed positions this can be a useful approach.
3. *Put a screen of high-permeability magnetic material in a position which screens circuit 2 from the flux of circuit 1.* The special alloy known as mumetal ($\mu_r \approx 10^5$) is used for this purpose. The method works only at low frequencies.

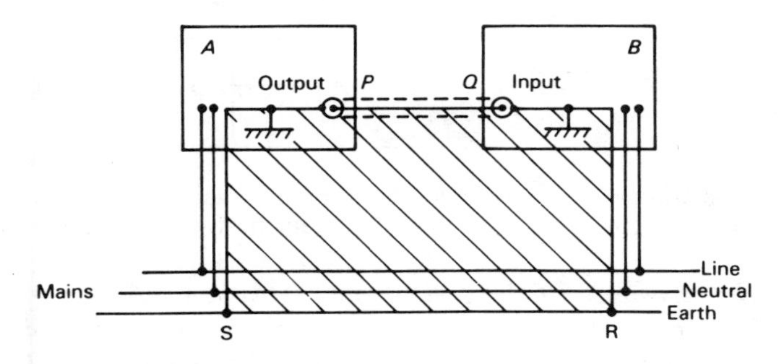

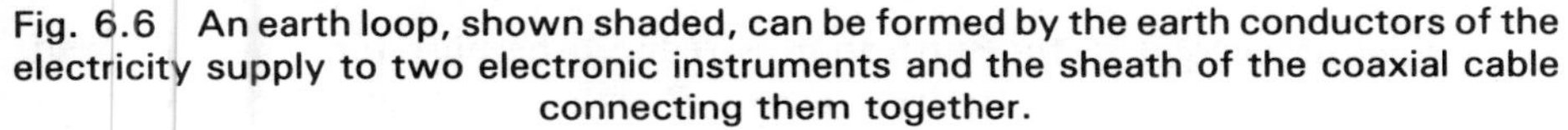
Fig. 6.6 An earth loop, shown shaded, can be formed by the earth conductors of the electricity supply to two electronic instruments and the sheath of the coaxial cable connecting them together.

A particularly troublesome type of electromagnetic interference is caused by **earth loops**. Figure 6.6 shows a typical situation in which this problem occurs. Two electronic instruments, A and B, are connected to the main supply by three-core cables whose earth conductors are connected to the instrument frames. The instruments are also connected to each other by a coaxial cable whose sheath is connected to the frames of the instruments to screen the signal wire from capacitively coupled interference. This arrangement produces a closed earth loop as shown by the cross-hatching in Fig. 6.6. The loop has low resistance and large currents are induced in it by the flux of the mains or some other source of interference. The induced current (I_i) flows through the resistance of the sheath of the coaxial cable (R_C) as shown in Fig. 6.7. This produces a spurious potential difference between P and Q which is added to the signal voltage V_A. Two solutions to this problem are possible:

1. *Break the earth loop so that no current can flow in it.* This is not as simple as it seems. Disconnecting the earth lead from the plug at R or S is potentially very dangerous because lethal voltages could appear on the instrument frames under fault conditions. This solution is only possible if a permanent earth connection between A and B can be ensured, for example by bolting them to the same rack. Ideally there should only be one path to earth from any point on the equipment. The alternative of disconnecting the cable sheath at P means that the signal earth follows the path $PSRQ$. The mutual inductance between this circuit and the mains is high.

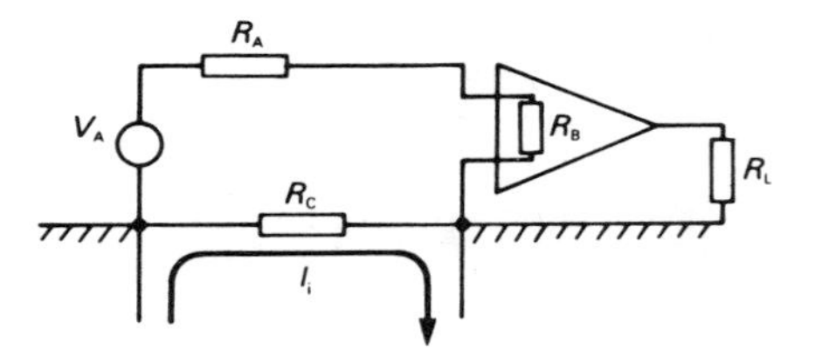

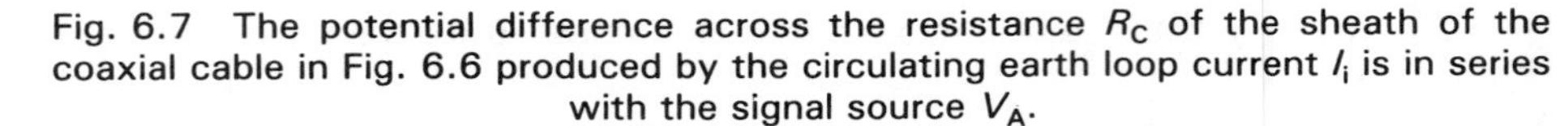

Fig. 6.7 The potential difference across the resistance R_C of the sheath of the coaxial cable in Fig. 6.6 produced by the circulating earth loop current I_i is in series with the signal source V_A.

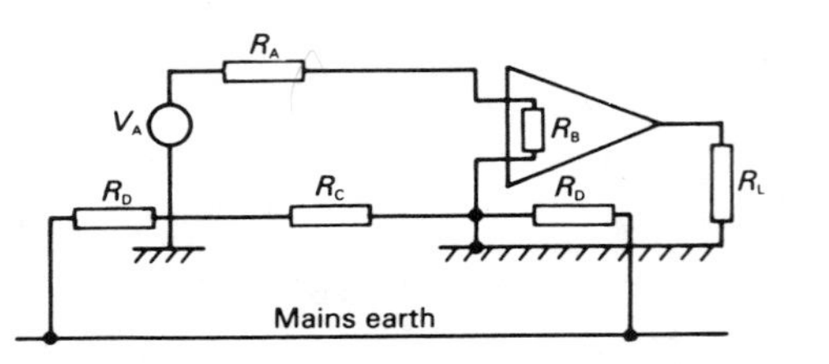

Fig. 6.8 The resistance of the earth loop can be increased by connecting resistors R_D between the signal earth and the supply earth of electronic instruments.

2. *Increase the resistance of the earth loop.* This can be achieved by putting resistors between the signal earth and the frame of each instrument as shown in Fig. 6.8. If the added resistors (R_D) are much larger than R_C, then the circulating current in the earth loop is reduced because the e.m.f. is fixed. The unwanted voltage appearing across R_C is reduced by a factor which is approximately $R_C/2R_D$. It is important that R_D is small enough so that no part of the system which can be touched reaches a lethal voltage when the maximum fault current is flowing. If $R_C = 0.1\ \Omega$ and $R_D = 50\ \Omega$, then the interference voltage is reduced by a factor of 1000. If, also, the mains fuse of each instrument is rated at 250 mA, then the voltage at P or Q cannot exceed 12.5 V with respect to earth.

Calculation of inductance

Before proceeding to a discussion of the ways of calculating inductance it is useful to review the situations in which such a calculation might be necessary. They are:

- To find the self- and mutual inductances of circuit components such as inductors and transformers.
- To find the inductance per unit length of two-wire transmission lines.
- To estimate the stray mutual inductance between parts of a circuit such as parallel tracks on a printed circuit board.

Inductance is calculated either by assuming a distribution of currents and computing the flux linkage, or by assuming a flux distribution and then finding the ampere turns needed to produce it. The first approach is best for many purposes, but we shall see later in the chapter that the two methods lead to upper and lower bounds for the inductance when energy methods are used. The calculation of inductance in simple cases is illustrated in the worked examples which follow.

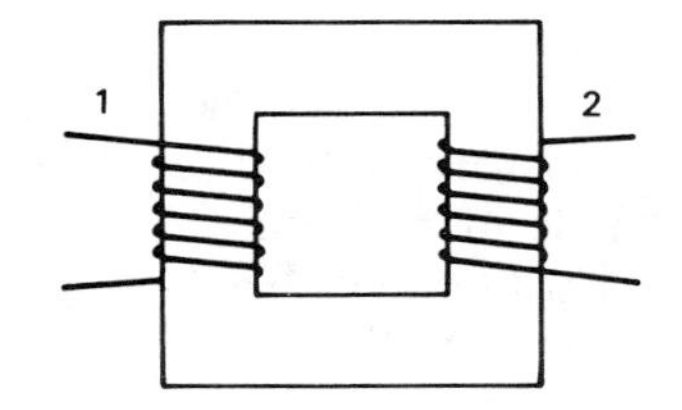

Fig. 6.9 The arrangement of a simple transformer.

Worked Example 6.1

Figure 6.9 shows an iron core which may be assumed to have a constant relative permeability of 1000, a cross-sectional area of 100 mm^2, and a mean path length around the magnetic circuit of 150 mm. Two coils of wire are wound on the core to form a simple transformer. Coil 1 has 100 turns and coil 2 1000 turns. Calculate the self-inductances of the two windings and the mutual inductance between them.

Solution Assume that a current of 1 A flows in coil 1. The magnetomotive force is 100 A turns. Applying the magnetic circuit law

$$H = 100/0.15 = 667 \text{ A m}^{-1}$$

Then $\Phi = 10^{-4}B = 10^{-4}\,\mu\text{H} = 8.4 \times 10^{-5}$ Wb

$$\Lambda_1 = 100 \times \Phi = 8.4 \times 10^{-3} \text{ Wb}$$

and

$$\Lambda_2 = 1000 \times \Phi = 84.0 \times 10^{-3} \text{ Wb}$$

so that

$$L_1 = 8.4 \text{ mH} \qquad \text{and} \qquad M = 84 \text{ mH}$$

If now a current of 1 A is assumed to flow in coil 2 we get

$$H = 6667 \text{ A m}^{-2} \qquad \Phi = 8.4 \times 10^{-4} \text{ Wb}$$
$$\Lambda_1 = 0.084 \qquad \text{and} \qquad \Lambda_2 = 0.84$$

so that

$$L_2 = 840 \text{ mH} \qquad \text{and} \qquad M = 84 \text{ mH}$$

We have shown, incidentally, that Equation (6.22) is valid for this case and that $M^2 = L_1L_2$, as would be expected from Equation (6.28) for the perfect coupling between the coils which has been assumed here.

Transformers are not only used for power supplies. They have an important function in matching impedances at audio frequencies and above. For further information on this subject, see Grossner.*

n real transformers there is always some leakage of flux, so that the calculation ust made is only approximate. To achieve greater accuracy it would be necessary o compute the detailed distribution of flux in and around the core. In practice, t is more likely that approximate calculations would be followed by empirical djustments to get the design right. The effects of leakage can be minimized by sing multilayer windings. At frequencies above 50 Hz the capacitance between he windings can be important because it limits the band of frequencies over which the transformer will work. The complete transformer can be represented by an equivalent circuit and methods exist for calculating the parameters of the circuit for any particular design of transformer. The core can be regarded as linear to a first approximation if it is made of a soft magnetic material and operated well below saturation. Inductors ('chokes') only have a single winding. They are sometimes made with an air gap in the magnetic circuit, so that the characteristic of the circuit is dominated by that of the gap and is, therefore, linear.

Worked Example 6.2

A coaxial line has inner and outer conductors with radii a and b respectively. Find an expression for the inductance per unit length of the line.

Solution The two conductors form a go-and-return circuit so that the currents in them are equal and opposite. The magnetic circuit law shows that there is a magnetic field only in the space between the conductors. Assuming that the current in the conductors is I, the magnetic circuit law shows that the field strength at a point between the conductors at radius r is

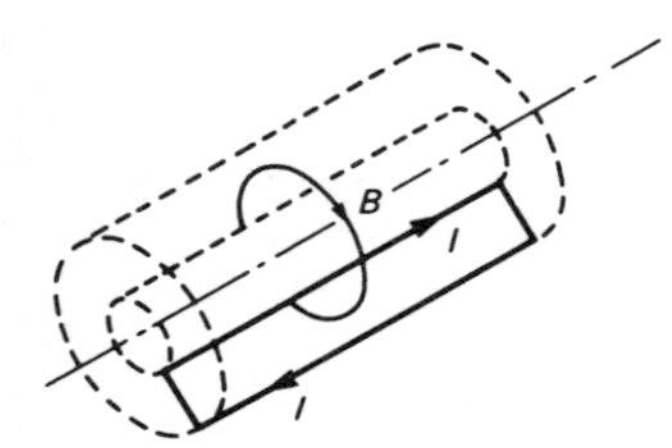

$$H = I/2\pi r$$

From the symmetry of the problem, this field is in the tangential direction. The conductors form a circuit having only a single turn, so the flux linked to it is found by calculating the flux crossing a unit length of a radial plane, as shown in the figure in the margin. Thus

$$\Lambda = \int_a^b \frac{\mu_0 I}{2\pi r}\,\mathrm{d}r = (\mu_0 I/2\pi)\ln(b/a)$$

The inductance per unit length is

$$L = \Lambda/I = (\mu_0/2\pi)\ln(b/a)$$

The significance of this result will become apparent in Chapter 7 when the use of coaxial cables as transmission lines is discussed. The estimation of the inductance per unit length of transmission lines of other shapes can be achieved by energy methods. This topic is treated later in this chapter.

Worked Example 6.3

Figure 6.10 shows a cross-section of adjacent tracks on a printed circuit board. Given that the tracks run parallel to each other for 50 mm, estimate the mutual inductance between the circuit comprising A and B and that comprising C and D.

Solution Take the origin of coordinates to be mid-way between A and B. Assume that equal and opposite currents I flow in A and B. Then the net flux density at a point lying between C and D and x from the origin is

The calculation of inductance in this and similar situations is discussed by Walker.*

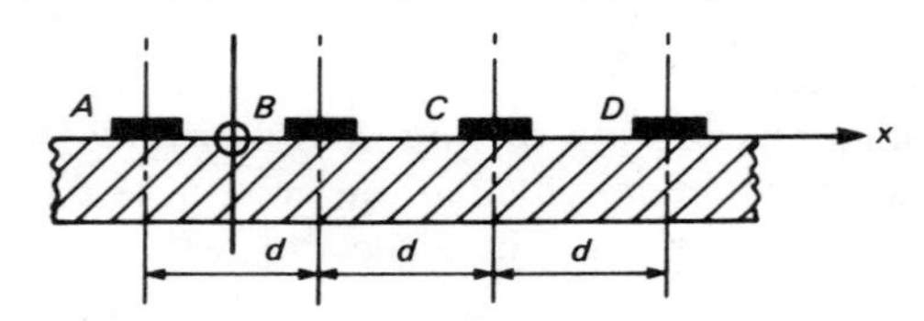

Fig. 6.10 A cross-sectional view of the conductors on a printed circuit board.

$$B = \frac{\mu_0 I}{2\pi}\left(\frac{1}{x - \frac{1}{2}d} - \frac{1}{x + \frac{1}{2}d}\right)$$

by applying the magnetic circuit law to A and B separately and then superimposing the results. The magnetic flux is normal to the plane containing the conductors at their mutual intersection. The flux due to A and B which is linked to the second circuit is found by integrating B over the area of circuit CD. Since the circuit has only one turn the result is

$$\Lambda = \frac{\mu_0 Il}{2\pi}\int_{3d/2}^{5d/2}\left(\frac{1}{x - \frac{1}{2}d} - \frac{1}{x + \frac{1}{2}d}\right)\mathrm{d}x$$

$$= \frac{\mu_0 Il}{2\pi}\ln\left(\frac{4}{3}\right)$$

The integral is taken between the mid planes of C and D as an approximation to the flux linkage to the real distributed currents in the conductors.

where l is the length over which the two circuits are coupled by the magnetic field. Putting in the numbers shows that the mutual inductance between the circuits is 2.8 nH. Whether this mutual inductance will lead to appreciable crosstalk between the circuits of the kind illustrated by Fig. 6.5 depends upon such factors as their impedances and signal levels. Note that in solving the problem we have assumed that the source currents can be regarded as line currents and that the dimensions of the conductors can be neglected. These assumptions are not strictly valid, but the result obtained by making them has an accuracy which is quite adequate for most purposes. In cases like this the real question is whether the mutual inductance is likely to be big enough to cause trouble.

Energy storage in the magnetic field

In Chapter 2 we saw that the work done in setting up an electric field could be thought of as being stored either in a lumped manner in the capacitance of the system ($\frac{1}{2}CV^2$) or distributed throughout space with an energy density $\frac{1}{2}\boldsymbol{D}\cdot\boldsymbol{E}$. In just the same way the work done in setting up a magnetic field can be regarded as being stored either lumped in the circuit inductance or distributed throughout the magnetic field.

It is well known from elementary circuit theory that the energy stored in an inductor is given by

$$W = \int_0^I Li\,\mathrm{d}i = \tfrac{1}{2}LI^2 \tag{6.30}$$

The proof of this equation is given by Compton.*

An alternative expression for W is obtained by substituting Λ/I for L

$$W = \tfrac{1}{2}\Lambda I \tag{6.31}$$

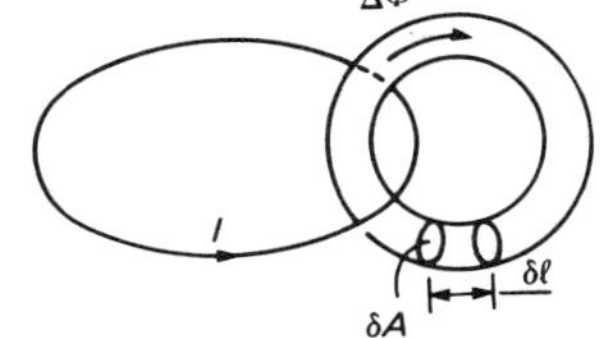

To derive an expression for the energy density in a magnetic field we consider a current loop and a typical flux tube linked to it, as shown in the figure in the margin. The stored energy associated with the tube, obtained from Equation (6.31), is

$$\Delta W = \tfrac{1}{2}\Delta\Phi I \tag{6.32}$$

Now if the cross-sectional area of the flux tube at a particular point is δA and the flux density at the same point is $\boldsymbol{B}$, then

$$\Delta\Phi = \boldsymbol{B}\cdot\delta A \tag{6.33}$$

Applying the magnetic circuit law gives

$$I = \oint \boldsymbol{H}\cdot \mathbf{d}\boldsymbol{l} \tag{6.34}$$

which can be approximated by $\Sigma \boldsymbol{H}\cdot\delta\boldsymbol{l}$ if the flux tube is considered to be made up of volume elements of length $\delta\boldsymbol{l}$ as shown in the figure in the margin (page 101). Substituting in Equation (6.32) for $\Delta\Phi$ and I produces

$$\Delta W = \tfrac{1}{2}(B\,\delta A)\Sigma(H\,\delta l) = \tfrac{1}{2}\Sigma(B\,\delta A)(H\,\delta l) \tag{6.35}$$

where the vector dot products have been dropped because the vectors concerned are all parallel to each other from the definition of the flux tube. $B\,\delta A$ is a constant of the flux tube, so it is permissible to bring it inside the summation. Finally, rearranging the terms of Equation (6.35) and noting that the product $\delta A\ \delta l$ is just the volume, δv, of the element defined by them, the energy stored in the flux tube becomes

$$\Delta W = \Sigma\tfrac{1}{2}BH\,\delta v \tag{6.36}$$

The rigorous derivation of Equation (6.37) can be found in Bleaney and Bleaney.*

so that the energy density in the magnetic field is evidently $\frac{1}{2}BH$. As in the case of the electric field considered in Chapter 2, a more rigorous derivation shows that the energy density in the field is given in general by

$$w = \tfrac{1}{2}\boldsymbol{B}\cdot\boldsymbol{H} \tag{6.37}$$

so that the total energy stored may be written

$$W = \tfrac{1}{2}\iiint \boldsymbol{B}\cdot\boldsymbol{H}\,\mathrm{d}v \tag{6.38}$$

A simple example will show that Equations (6.30) and (6.38) give identical results for the energy stored in an inductor.

Worked Example 6.4

Figure 6.11 shows an iron ring of square cross-section. An inductor is made by winding N turns of wire uniformly on the ring as shown. Assuming that the permeability of the iron is constant, calculate the energy stored in the inductor when a current I is flowing in the winding, using Equations (6.30) and (6.38).

Solution (a) *Using Equation (6.30)*

When a current I is flowing in the winding the magnetic field at radius r is, by the magnetic circuit law,

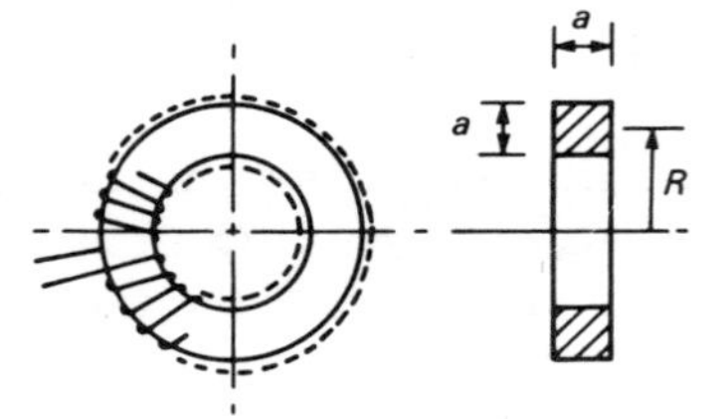

Fig. 6.11 The arrangement of a toroidal inductor.

$$H = NI/2\pi r$$

so the flux circulating within the iron is

$$\Phi = \frac{\mu NIa}{2\pi}\int_{R-\frac{1}{2}a}^{R+\frac{1}{2}a}\frac{\mathrm{d}r}{r}$$

$$= \frac{\mu NIa}{2\pi}\ln\left(\frac{R+\frac{1}{2}a}{R-\frac{1}{2}a}\right)$$

The inductance is

$$L = N\Phi/I$$

$$= \frac{\mu N^2 a}{2\pi}\ln\left(\frac{R+\frac{1}{2}a}{R-\frac{1}{2}a}\right)$$

and the stored energy is

$$W = \frac{\mu N^2 I^2 a}{4\pi}\ln\left(\frac{R+\frac{1}{2}a}{R-\frac{1}{2}a}\right)$$

(b) *Using Equation (6.38)*
Taking a volume element in the shape of a ring of radius r, thickness $\mathrm{d}r$ and width a, the stored energy in the ring is

$$\mathrm{d}W = \tfrac{1}{2}\mu H^2\, 2\pi ra\,\mathrm{d}r = \frac{\mu N^2 I^2 a\,\mathrm{d}r}{4\pi r}$$

Integrating this expression over r from $(R - \frac{1}{2}a)$ to $(R + \frac{1}{2}a)$ gives exactly the same result as before for the stored energy. It is important to realize that this calculation has been made possible by assuming that μ is a constant so that the system is linear. When the non-linearity of the iron has to be taken into account the inductance of the coil is not a constant and the energy stored depends upon the previous history of the iron.

Calculation of inductance by energy methods

Further information about the application of energy methods to the calculation of inductance is given by Hammond.*

We have seen that energy methods provide a useful way of obtaining quite good approximations to circuit parameters such as capacitance in cases which cannot be solved by analytical methods. These methods can also be used to estimate inductances. The argument runs exactly parallel to the cases of capacitance and resistance so it will not be given in detail. Starting from Equation (6.30) we discover that

$$L \leqslant 2W'/I^2 \tag{6.39}$$

where W' is the estimate of the stored energy produced by assuming that the current distribution is given. The corresponding lower bound can be obtained by making use of the definition of self-inductance to give another expression for the stored energy

$$W = \Lambda^2/2L \tag{6.40}$$

If an estimate W'' of the stored energy is obtained by assuming that the flux distribution is given, then

$$L \geqslant \Lambda^2/2W'' \tag{6.41}$$

The application of the method is illustrated in the example which follows.

Worked Example 6.5 Estimate the inductance per unit length of the coaxial line shown in Fig. 6.12.

This coaxial line is identical to that shown in Fig. 2.10. The reason for this choice will become apparent in the next section.

Solution (*a*) *Upper bound*

The upper bound is obtained by assuming an approximate current distribution. One possibility, as shown in Fig. 6.12(b), is to distribute the current uniformly over the broad faces of the outer conductor and over the whole of the inner conductor. The approximate magnetic equipotential surfaces associated with this distribution of current are also shown in Fig. 6.12(b). Now the inductance of the region between each pair of equipotentials is associated with part of the total current flow. The total inductance is, therefore, approximately that of all such elements in parallel with each other.

The reciprocal of the inductance of an element subtending an angle $d\theta$ is obtained from Worked Example 6.2:

$$d\left(\frac{1}{L}\right) = d\theta/(\mu_0 \ln(b/a))$$

Since the outer conductor is twice the length of the inner conductor it follows that $b/a = 2$ for all the elements. The above equation is integrated over all the strips to give

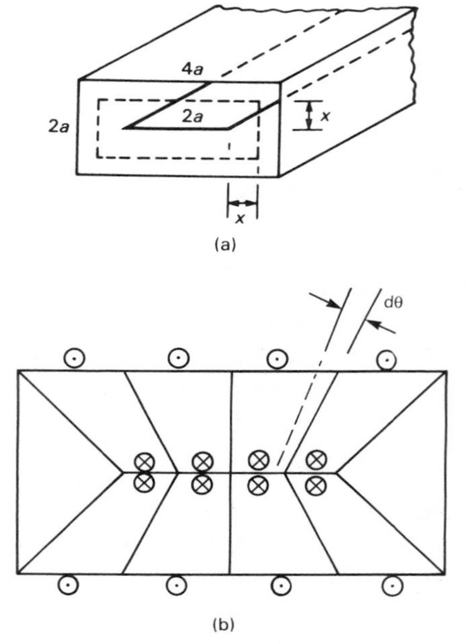

Fig. 6.12 A rectangular coaxial line with a strip centre conductor.

$$L \leqslant \mu_0(\ln 2)/\pi = 0.22\mu_0$$

ɩs the upper bound of L. It will be noticed that the triangular regions at each side ›f the diagram have been omitted from the calculations. The inductances of these egions are large because the ratio $b/a \rightarrow \infty$. Thus these inductances have the ·ffect of reducing the estimate of L slightly and so the figure given above is :ertainly an upper bound on L.

b) Lower bound

Го obtain the lower bound we assume approximate flux lines as shown in ᵎig. 6.12(a). Applying the magnetic circuit law to the path shown by the broken ine and assuming that H is everywhere parallel to that line gives

$$H = I/(4a + 8x)$$

Гhe flux contained in a unit length of tube of thickness dx is

$$\Phi = \mu_0 H\mathrm{d}x$$

$$= \frac{\mu_0 I\mathrm{d}x}{(4a + 8x)}$$

.o the self-inductance of the element is

$$\mathrm{d}L = \frac{\mu_0 \mathrm{d}x}{(4a + 8x)}$$

Гhe inductances of the elements are effectively in series with each other so the total elf-inductance is obtained by integrating this expression from $x = 0$ to $x = a$. Гhe result is

$$L \geqslant (\mu_0/8)\ln 3 = 0.137\mu_0$$

Гhe best estimate of L is given by the average of the upper and lower bounds, so

$$L = (0.18 \pm 0.04)\mu_0$$

Гhe exact figure is $0.17\mu_0$ so the approximate value is in error by about 6%.

Гhe method can also be applied to problems involving magnetic circuits to give nore accurate results than those obtained by the elementary method illustrated in Vorked Example 6.1.

Гhe *LCRZ* analogy

ᐠ comparison between Worked Examples 2.6 and 6.5 shows a marked resem-›lance. This suggests that, if we can calculate the capacitance for a particular ɩrrangement of electrodes, we should be able to deduce an inductance for the same ͺeometry from it. The justification for this is that the field patterns in each case nust satisfy Laplace's equation (1.27). Since this also applies to current flowing n a conductor, we can add resistance to the discussion. We will assume that the nedium between the electrodes has the same properties everywhere.

Consider, for example, the coaxial geometry shown in Fig. 6.13(a). We have ;hown that the capacitance per unit length is

$$C = 2\pi\epsilon/\ln(b/a) \tag{6.42}$$

(see Worked Example 2.3). Similarly, a simple extension of Worked Example 6.2 gives the inductance per unit length as

$$L = \mu \ln(b/a)/2\pi \tag{6.43}$$

It is easy to show that the resistance per unit length for radial current flow is given by

$$R = \rho \ln(b/a)/2\pi \tag{6.44}$$

Similarly, for the parallel strip geometry shown in Fig. 6.13(b) we obtain the following parameters per unit length if fringing fields are neglected:

$$C = \epsilon w/d \tag{6.45}$$

$$L = \mu d/w \tag{6.46}$$

$$R = \rho d/w \tag{6.47}$$

We deduce that for all such two-dimensional arrangements of electrodes the parameters per unit length are given by

$$C = \epsilon\Gamma \tag{6.48}$$

$$L = \mu/\Gamma \tag{6.49}$$

$$R = \rho/\Gamma \tag{6.50}$$

where the geometry of the system determines the parameter Γ. For coaxial geometry

$$\Gamma = 2\pi/\ln(b/a) \tag{6.51}$$

and for parallel strips

$$\Gamma = w/d \tag{6.52}$$

By multiplying Equations (6.48) and (6.49) together we obtain

$$LC = \epsilon\mu \tag{6.53}$$

which is a constant whose value depends only on the properties of the material surrounding the electrodes. Dividing Equation (6.49) by Equation (6.48) we get

$$\sqrt{L/C} = \sqrt{\mu/\epsilon}/\Gamma = Z/\Gamma \tag{6.54}$$

where Z has the dimensions of resistance and depends only upon the properties of the material. Equations (6.48)–(6.50) can be used to find any two of the trio

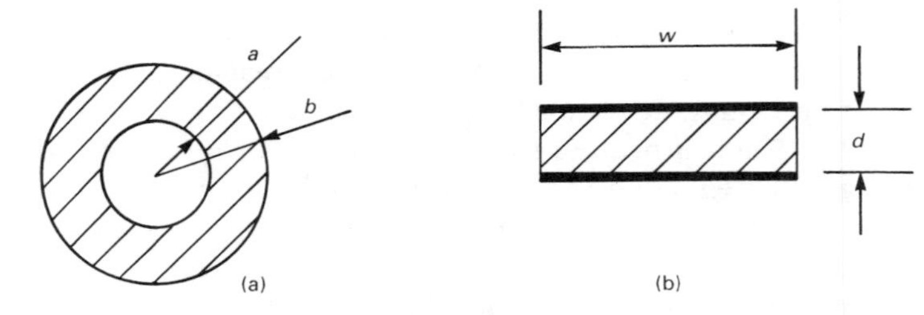

Fig. 6.13 (a) Coaxial and (b) parallel-strip geometries.

of parameters C, L and R for a particular geometry once one of them is known. Note carefully that the magnetic flux lines in each case are at right angles to the electric field lines. The significance of this will become apparent in Chapter 7.

For further discussion of the *LCRZ* analogy and its applications see Walker.*

These ideas can be extended further by invoking the principle of duality discussed on p. 50. It is easy to see from Fig. 6.13(b) that, when the roles of the field and equipotential lines are exchanged the parameters of the dual system are given by

$$C' = \epsilon/\Gamma \tag{6.55}$$

$$L' = \mu\Gamma \tag{6.56}$$

$$R' = \rho\Gamma \tag{6.57}$$

Worked Example 6.6

Show how the *LCRZ* analogy can be used to deduce the results of Worked Examples 2.6 and 6.5 from each other.

Solution In Worked Example 2.6 it was shown that

$$\pi\epsilon_0/\ln 2 \leqslant C \leqslant 8\epsilon_0/\ln 3$$

so the geometrical parameter must lie in the range

$$\pi/\ln 2 \leqslant \Gamma \leqslant 8/\ln 3$$

and, from Equation (6.49), the inductance must lie in the range

$$(\mu_0/8)\ln 3 \leqslant L \leqslant (\mu_0/\pi)\ln 2$$

which agrees exactly with the results of Worked Example 6.5.

Energy storage in iron

So far our discussion of energy in magnetic fields has been restricted to those cases where the permeability is constant. To investigate storage of energy when the permeability is not constant, we consider an iron ring of mean circumference l and cross-sectional area A which has N turns of wire wound upon it, as shown in Fig. 6.11. In time $\mathrm{d}t$ let the current in the winding increase from I to $(I + dI)$ and the flux in the iron increase from Φ to $(\Phi + d\Phi)$. The induced e.m.f. in the winding is $N\,\mathrm{d}\Phi/\mathrm{d}t$, so the work done in the time interval is

$$\delta W = NI\frac{\mathrm{d}\Phi}{\mathrm{d}t}\,\mathrm{d}t = NI\,\mathrm{d}\Phi \tag{6.58}$$

But

$$\Phi = BA \qquad \text{and} \qquad Hl = NI$$

so

$$\mathrm{d}W = AlH\,\mathrm{d}B$$

The change in the stored energy when the system is taken from flux density B_1 to B_2 is given by

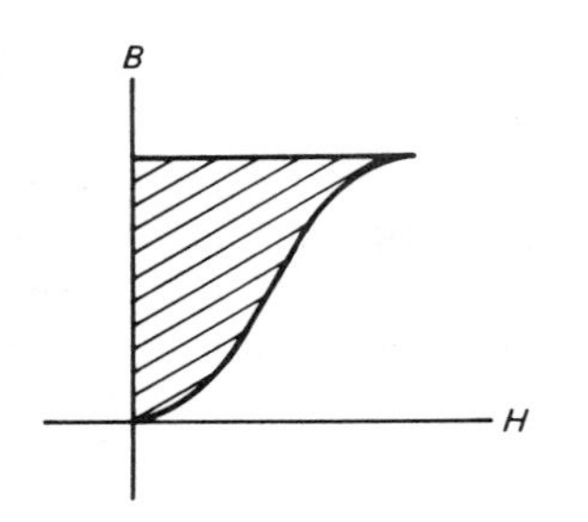

Fig. 6.14 The work done in magnetizing a piece of iron is represented by the area between the initial magnetization curve and the vertical axis.

$$W = Al \int_{B_1}^{B_2} H \, \mathrm{d}B \tag{6.59}$$

But Al is just the volume of the iron, so the change of energy density is

$$w = \int_{B_1}^{B_2} H \, \mathrm{d}B \tag{6.60}$$

The rigorous proof of this expression is given by Reitz and Milford.* It is important to remember that the process of magnetization is not reversible in the thermodynamic sense. That is, not all the energy stored during the process of magnetization can be recovered.

Once again more rigorous argument shows that Equation (6.60) is correct for all cases if the scalar product $H\,\mathrm{d}B$ is replaced by a vector dot product $\boldsymbol{H}\cdot\mathbf{d}\boldsymbol{B}$. The work done in magnetizing an initially unmagnetized specimen of iron is illustrated by Fig. 6.14. The integral in Equation (6.60) is represented by the shaded area between the initial magnetization curve and the vertical axis.

Hysteresis loss

A case of particular interest is the work done in taking a piece of iron once around its hysteresis loop. In any iron-cored inductor or transformer the core is taken through this cycle for every cycle of the current in the windings. Figure 6.15 shows a typical hysteresis loop. The change in stored energy per unit volume in going from 1 to 2 is represented by the shaded area in Fig. 6.16(a). Let us denote this energy by w_{12}. The shaded area in Fig. 6.16(b) represents the change in stored energy per unit volume when the iron is taken from 3 to 2, a change which is mathematically possible but physically impossible. This energy is w_{32}. The real physical

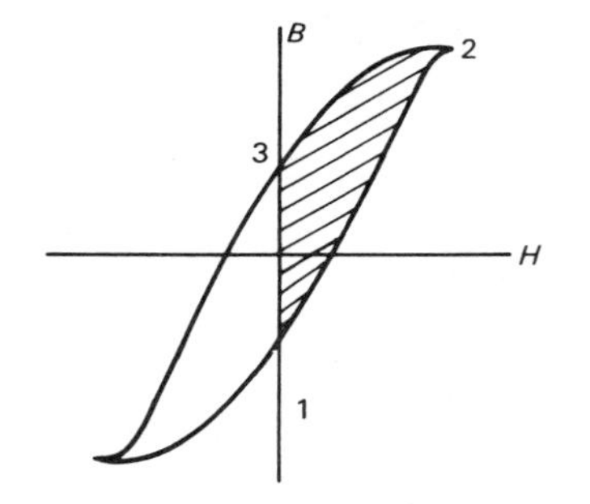

Fig. 6.15 The work done in taking a piece of iron around its hysteresis loop is proportional to the area of the loop.

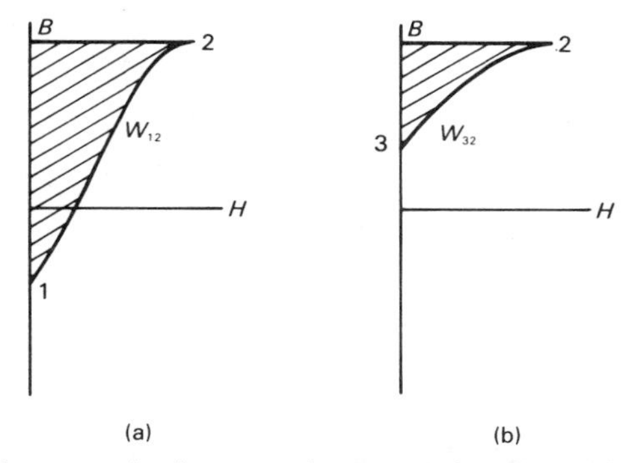

Fig. 6.16 (a) The increase in the stored energy of a piece of iron when it is taken around its hysteresis loop from 1 to 2; (b) the energy recovered when it is taken from 2 to 3.

change is obtained by exchanging the limits of integration in Equation (6.60) so that $w_{23} = -w_{32}$. The change in energy density resulting from going from 1 to 3 via 2 is

$$w_{13} = w_{12} - w_{32}$$

which is represented by the difference between the areas shown in Figs 6.16(a) and 6.16(b), and by the shaded area in Fig. 6.15. Since the hysteresis loop is symmetrical the work done in encircling the loop once is equal to the area of the loop multiplied by the volume of the iron. It is evident that this work is not zero since all real ferromagnetic materials have hysteresis loops with non-zero areas. This phenomenon means that electrical energy is changed into heat in the iron core at a rate proportional to the frequency of the signal applied to the device. The conversion of energy in this way is known as **hysteresis loss**. The loss is minimized by using soft magnetic materials for transformer cores, as an examination of Fig. 5.13 will show.

Eddy currents

When any conducting circuit is placed in a changing magnetic field currents are induced in the circuit, as we saw earlier. This remains true when the circuit is a solid piece of metal. In that case currents, known as **eddy currents**, circulate within the metal, causing power loss by ohmic heating. They are put to use in the industrial eddy-current heating (ECH) process, which is used for brazing metal components together. In other cases, particularly in transformer cores, they give rise to unwanted losses. The full theory of the generation of eddy currents is beyond the scope of this book, but it is possible to give an approximate treatment of the special case in which the conductor is in the form of a thin strip with the magnetic field parallel to its length.

Figure 6.17 shows a section of a strip of conducting material whose width (l) is much greater than its thickness ($2t$). The strip is placed in a uniform magnetic field directed along the strip with flux density $B_0 \cos(\omega t)$. The eddy currents in the strip will flow in closed loops like the one shown. From Faraday's law we know that the e.m.f. induced in the loop is given approximately by

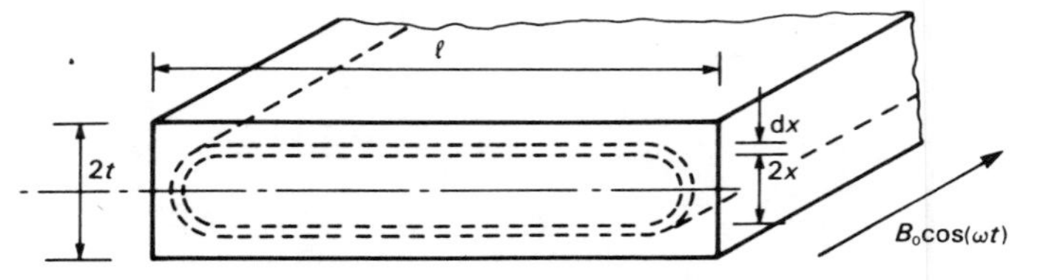

Fig. 6.17 The element of volume used in estimating the eddy-current loss in a thin conducting strip.

$$\mathcal{E} = -2lx\frac{\mathrm{d}B}{\mathrm{d}t} = 2lx\omega B_0 \sin(\omega t) \tag{6.61}$$

since the area of the loop is approximately $2lx$.

The resistance of the loop is approximately

$$R = 2\rho l/\mathrm{d}x \tag{6.62}$$

per unit length of the strip, where ρ is the resistivity of the material.

The mean power loss in the loop is

$$\mathrm{d}W = |\mathcal{E}|^2/2R = (l\omega^2 B_0^2/\rho)\, x^2\,\mathrm{d}x$$

so that the mean power loss per unit length of the strip is

$$W = (l\omega^2 B_0^2/\rho)\int_0^t x^2\,\mathrm{d}x$$
$$= \frac{lt^3\omega^2 B_0^2}{3\rho} \tag{6.63}$$

We can also express this as mean power loss per unit surface area

$$W_s = t^3\omega^2 B_0^2/3\rho \tag{6.64}$$

In deriving this expression we have implicitly assumed that the eddy currents are not strong enough to have a significant effect upon the strength of the magnetic field. Equation (6.64) shows that the eddy-current losses depend very strongly on the thickness of the strip. This is the reason why transformer cores are made of thin strips or **laminations** of steel which are insulated from each other by a coating of lacquer. They are normally made of a special steel which has a high resistivity because, as can be seen from Equation (6.64), this also helps to reduce the losses. The mean power loss also increases rapidly with increasing frequency. At radio-frequencies the losses in laminated iron cores are unacceptedly high and magnetic oxides of iron known as **ferrites** are used instead because they have much higher resistivities.

Real electronic components

For further discussion of these subjects, consult Sangwine,* Walker* and Grossner.*

By this stage it should be apparent to the reader that the behaviour of real electronic components is much more complicated than that of the lumped resistors, capacitors and inductors used in circuit theory. Whenever a current flows through a component a magnetic field will be generated, so resistors and capacitors must have some self-inductance. The resistance in the winding of an inductor or transformer means that there must be some voltage drop across the device even under d.c. conditions. The existence of a potential difference between the different parts of the winding implies that there is some capacitance present. Although these parasitic effects can usually be neglected at low frequencies, they must be included in the circuit representations of the components at high frequencies.

Summary

In this chapter Faraday's law of electromagnetic induction has been introduced. It has been shown that an electromotive force can be induced in a circuit which has magnetic flux linked to it either by changing the area of the circuit or by changing the strength of the magnetic field. From this starting point the idea of inductance was introduced, so establishing another link between electric circuit parameters and electromagnetic field theory. Stray inductances between parts of electronic circuits can produce unwanted coupling. The causes of electromagnetic interference were considered and some cures suggested. The calculation of inductance was discussed with a number of worked examples to illustrate both direct calculation and the use of energy methods. The latter are based on the idea of the stored energy associated with a magnetic field being distributed throughout the field. The analogy between L, C and R in two-dimensional problems was introduced as a way of deducing these parameters from each other. The special case of energy storage in iron was discussed, leading to an expression for the work done in taking a sample of iron once around its hysteresis loop and to the idea of hysteresis loss. Finally, the consideration of the effects of a changing magnetic field on a solid conductor led to a discussion of eddy currents and of the losses associated with them.

Problems

6.1 Two coaxial circular wire loops, with radii 0.01 m and 0.5 m, lie in the same plane. Estimate their mutual inductance. If the smaller loop is rotated about a diameter, how does the mutual inductance depend upon the angle between the planes of the two loops?

6.2 Figure 6.18 shows the arrangement of a transformer used to measure the

This type of transformer is known as a **current transformer**. It is used as a way of monitoring the currents flowing in radar pulse modulators. A variant form is the clip-on ammeter. In this instrument the secondary coil is concentrated on one part of the core and the core itself is hinged so that it can be clipped over the primary conductor.

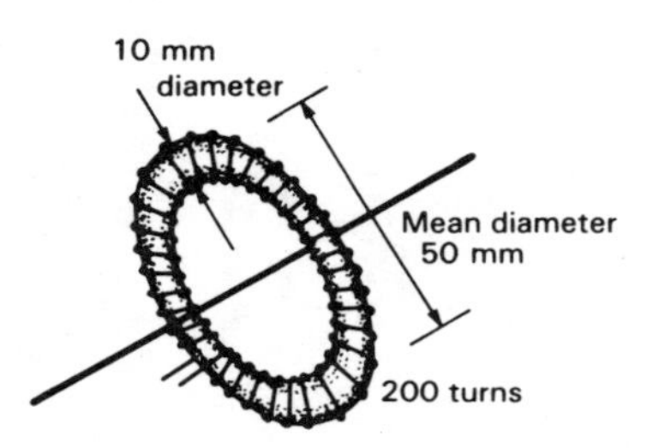

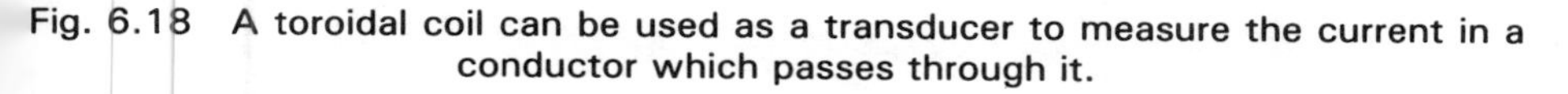
Fig. 6.18 A toroidal coil can be used as a transducer to measure the current in a conductor which passes through it.

current in a high-voltage cable. Calculate the open circuit voltage induced in the toroidal winding if the material of the toroid has a relative permeability of 130 and the current in the central conductor is 10 A rms at 1 kHz. Does this answer change if the primary conductor is not on the axis of the transformer?

6.3 The centres of the line and neutral conductors of a 30 A ring main cable are 5 mm apart. Estimate the maximum e.m.f. induced in a circuit enclosing an area of 0.1 m^2 which is 2 m from the cable.

6.4 Figure 6.19 shows the arrangement of a ferrite pot core inductor for use at radio-frequencies. Given that the relative permeability of the ferrite is 130, what is the inductance of the coil? If the ferrite begins to saturate at a flux density of 0.15 T, what is the maximum current in the coil for linear operation?

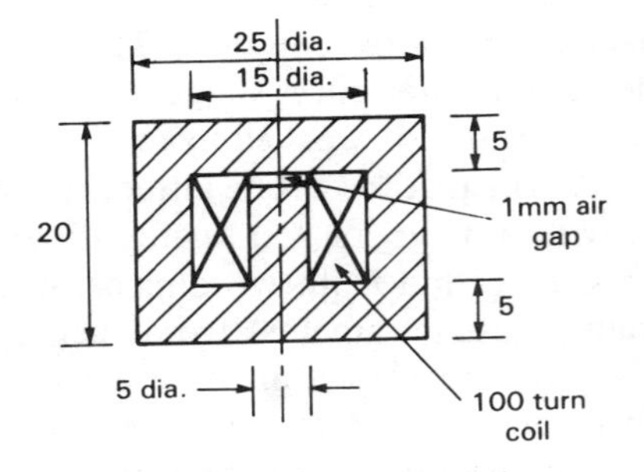

Fig. 6.19 A cross-sectional view of an inductor made with a cylindrical pot core of magnetic material.

6.5 A toroidal transformer core is made by winding 20 turns of Permalloy D strip 10 mm wide and 0.5 mm thick onto a 40 mm diameter former. A toroidal primary winding of 100 turns is wound uniformly over the core and a secondary winding of 500 turns is wound uniformly over the primary. Calculate the self-inductances of the two windings and the mutual inductance between them, given that Permalloy D has a relative permeability of 4000.

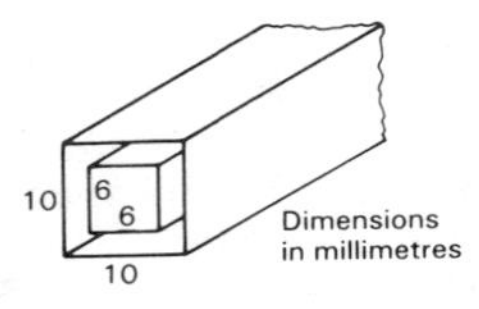

6.6 Estimate the inductance per unit length of the square coaxial arrangement of conductors shown in the figure in the margin.

Transmission lines 7

Objectives

- ☐ To introduce electromagnetic waves through a discussion of transmission lines as distributed circuits.
- ☐ To discuss the solutions to the wave equation for sinusoidal waves and pulses, and to introduce the idea of phase velocity.
- ☐ To explain the concept of characteristic impedance, and its relationship to the power flow in the line.
- ☐ To explain the use of complex notation to describe waves.
- ☐ To show that a wave is reflected at the end of a line unless it is terminated by a load equal to its characteristic impedance.
- ☐ To introduce the terms 'voltage reflection coefficient' and 'voltage standing wave ratio'.
- ☐ To show how impedances are transformed by a transmission line, and to introduce the idea of the quarter-wave transformer as a matching device.
- ☐ To consider the field description of a transmission line and to show how the phase velocity and characteristic impedance can be calculated.
- ☐ To derive expressions for the electric and magnetic fields in a coaxial cable and to consider how the power flow in the line may be calculated from them.

Two-wire transmission lines appeared early in the history of the practical application of electricity with the advent of the electric telegraph. They are still an important means of transmitting both electrical power and information, although for the latter purpose they are now supplemented by radio, microwave links and optical fibres. The commonest forms are parallel-wire and twisted-pair lines, used for electricity distribution and telephone connections, and coaxial cables, used for television aerial downleads and for interconnecting electronic instruments.

Throughout this book I have tried to show the links between the field and the circuit approaches to the description of electromagnetic phenomena. These provide alternative ways of dealing with problems. Electronic engineers are generally happiest when they are able to use circuit methods, with the properties of the components represented by equivalent circuits. Field theory provides an alternative to experimental measurements as a means of determining the equivalent circuit parameters. In this chapter, however, the field and circuit approaches draw even closer together. We shall see that the transmission of signals on transmission lines can be described in terms of propagation of electromagnetic waves. We shall also see that, when the dimensions of a circuit become comparable with the wavelength of these waves, elementary circuit theory breaks down.

The circuit theory of transmission lines

It was shown in Chapters 2 and 6 that a two-wire transmission line has shunt capacitance and series inductance uniformly distributed along its length. At low

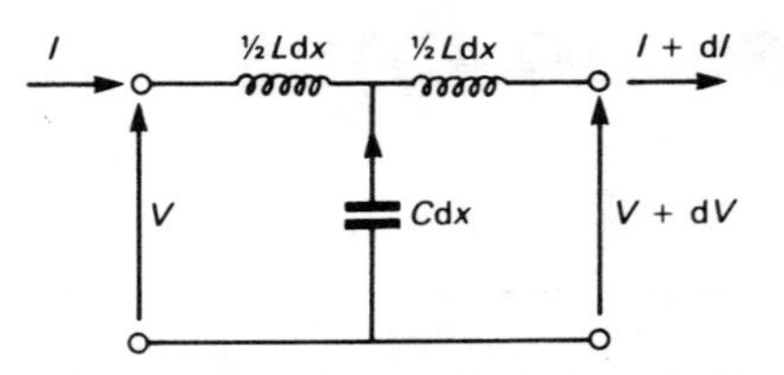

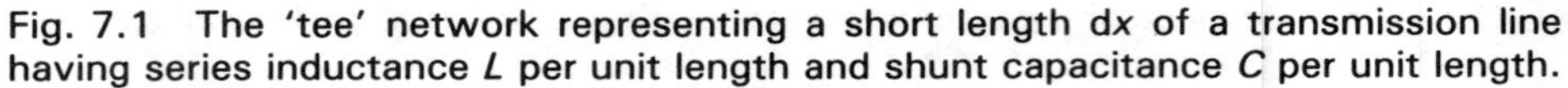

Fig. 7.1 The 'tee' network representing a short length dx of a transmission line having series inductance L per unit length and shunt capacitance C per unit length.

frequencies it is usually possible to ignore these impedances and treat circuits as sets of **lumped components** connected together by wires whose length has no effect upon the operation of the circuit. We must now investigate the circumstances in which this approximation is not valid and the consequences for the design of electronic circuits.

Figure 7.1 shows the circuit representation of an element of line of length dx. If the line has inductance L per unit length and capacitance C per unit length, then the inductance and capacitance of the element must be $L\,\mathrm{d}x$ and $C\,\mathrm{d}x$.

The inductance has been divided into two so that the network is a symmetrical **tee section**. For the moment we assume that the whole line stretches to infinity in either direction, being made up of identical tee sections joined together in a cascade. If the voltage across the capacitor is V', then

$$V' = V - \tfrac{1}{2}L\frac{\mathrm{d}I}{\mathrm{d}t}\mathrm{d}x = V + \mathrm{d}V + \tfrac{1}{2}L\frac{\mathrm{d}}{\mathrm{d}t}(I + \mathrm{d}I)\,\mathrm{d}x$$

When the second-order term $\dfrac{\mathrm{d}(\mathrm{d}I)}{\mathrm{d}t}$ is neglected we obtain

$$\frac{\partial V}{\partial x} = -L\frac{\partial I}{\partial t} \qquad (7.1)$$

The current through the capacitor is

$$\mathrm{d}I = -C\,\mathrm{d}x\frac{\mathrm{d}V'}{\mathrm{d}t}$$

To first order $V' = V$, so that

$$\frac{\partial I}{\partial x} = -C\frac{\partial V}{\partial t} \qquad (7.2)$$

This pair of equations is sometimes refered to as the **telegraphers' equations**.

The pair of simultaneous differential equations (7.1) and (7.2) defines the relationship between the voltage and the current in the limit when the length (dx) of the element shrinks to zero and the line becomes completely uniform. The current can be eliminated from these equations by differentiating Equation (7.1) with respect to x, and Equation (7.2) with respect to t, since

$$\frac{\partial^2 I}{\partial x \partial t} = \frac{\partial^2 I}{\partial t \partial x}$$

giving

$$\frac{\partial^2 V}{\partial x^2} = LC\frac{\partial^2 V}{\partial t^2} \qquad (7.3)$$

Similarly, differentiating Equation (7.1) with respect to t and Equation (7.2) with respect to x, we get

$$\frac{\partial^2 I}{\partial x^2} = LC\frac{\partial^2 I}{\partial t^2} \tag{7.4}$$

Equations (7.3) and (7.4) are examples of the one-dimensional form of the **wave equation**. For sinusoidal waves their general solution has the form

For a more detailed treatment of the mathematics of waves consult a text such as Kreyszig, E., *Advanced Engineering Mathematics* (5th edn), Wiley (1983).

$$V = V_1 \cos(\omega t \pm kx) + V_2 \sin(\omega t \pm kx) \tag{7.5}$$

and

$$I = I_1 \cos(\omega t \pm kx) + I_2 \sin(\omega t \pm kx) \tag{7.6}$$

where V_1, V_2, I_1 and I_2 are constants whose values are determined by the boundary conditions of the problem. Substituting these expressions into the wave equations shows that they are acceptable solutions provided that

$$k^2 = \omega^2 LC \tag{7.7}$$

The physical significance of these solutions can be discussed, without loss of generality, by considering the behaviour of the function

$$V = V_1 \cos(\omega t - kx) \tag{7.8}$$

The form of this function is shown in Fig. 7.2. The solid curve shows the variation of voltage with position along the line when $t = 0$. At a later time $t = t_1$ the wave has moved to the position shown by the broken line. This can be verified by noting that the first zero of the function is given by

$$(\omega t - kx) = -\pi/2$$

or

$$x = (\pi/2k) + (\omega t/k) \tag{7.9}$$

This equation shows that the zero crossing, and hence the whole wave, is moving in the positive x-direction with a constant velocity given by

$$v_p = \omega/k \tag{7.10}$$

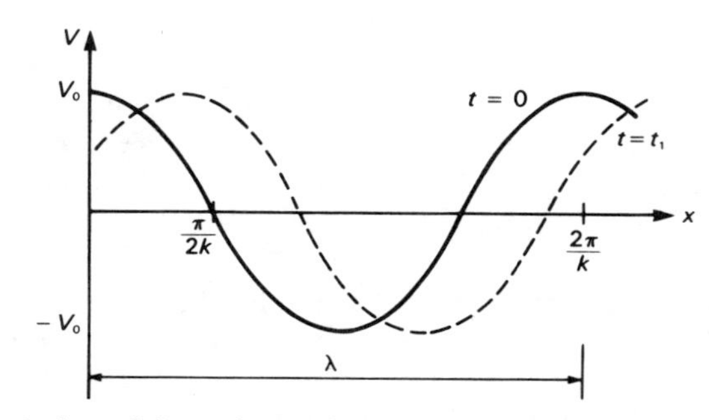

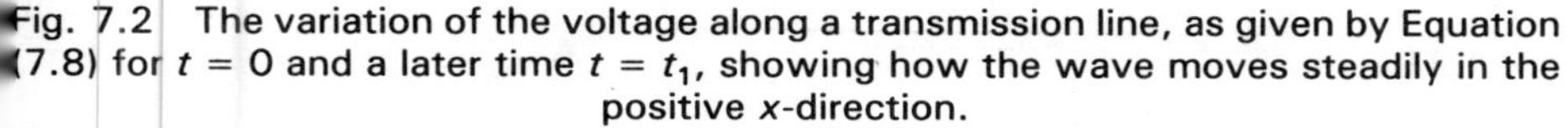
Fig. 7.2 The variation of the voltage along a transmission line, as given by Equation (7.8) for $t = 0$ and a later time $t = t_1$, showing how the wave moves steadily in the positive x-direction.

This velocity is known as the **phase velocity** of the wave. From Equation (7.7) we also have

$$v_p = 1/\sqrt{(LC)} \tag{7.11}$$

Equation (6.53) shows that v_p depends only on the properties of the material between the conductors and not on their geometry. The **propagation constant**, or **wavenumber**, k is related to the wavelength λ of the wave by

$$k = 2\pi/\lambda \tag{7.12}$$

as can be seen from Fig. 7.2. If k is negative the phase velocity is also negative and the wave is travelling in the negative x-direction. This corresponds to the positive signs in Equation (7.5).

The alternative solution, involving the sine function in place of the cosine, gives a wave which is shifted in phase by 90°. By making a suitable choice of the constants V_1 and V_2 in Equation (7.5) we can make the amplitude and phase of the wave what we will. This is made clearer by a consideration of the alternative form of (7.5):

$$V = V_0 \cos(\omega t \pm kx \pm \phi) \tag{7.13}$$

Where the amplitude V_0 and the phase ϕ are arbitrary constants whose values are to be determined by the boundary conditions. The equivalence of the two forms of the wave given in Equations (7.5) and (7.13) can be demonstrated by expanding the cosine function in Equation (7.13).

Representation of waves using complex numbers

In the theory of a.c. circuits 'j-notation' is an indispensable tool. Before proceeding to a discussion of the properties of transmission lines it is necessary to extend the notation to problems where the voltages and currents vary sinusoidally in distance as well as in time. To do this consider the voltage

For a discussion of the use of complex numbers in the analysis of a.c. circuits, see Carter, G.W. and Richardson, A., *Techniques of Circuit Analysis*, Cambridge University Press (1972).

$$V' = \mathrm{j}V_1 \sin(\omega t - kx) - \mathrm{j}V_2 \cos(\omega t - kx) \tag{7.14}$$

When this voltage is added to the general solution for the wave in the positive x direction, obtained from Equation (7.5) by taking the negative signs, the result is

$$\begin{aligned} V &= V_1(\cos(\omega t - kx) + \mathrm{j}\sin(\omega t - kx)) \\ &\quad - \mathrm{j}V_2(\cos(\omega t - kx) + \mathrm{j}\sin(\omega t - kx)) \\ &= \hat{V}\exp[\mathrm{j}(\omega t - kx)] \end{aligned} \tag{7.15}$$

where the amplitude of the wave $\hat{V}$ is, in general, complex. It turns out that Equation (7.15) provides a more convenient form of representation for a wave than Equation (7.5), but it must be remembered that it is only the real part of Equation (7.15) which has physical significance.

Hence, whenever the form in Equation (7.15) is used to represent a wave in a problem it is implicit that, when an expression has been obtained as the solution to the problem, the real part of it is to be taken as having physical significance. It is easy to verify by substitution that Equation (7.15) is a solution of Equation (7.3). The two arbitrary constants in Equation (7.5) are present as the real and imaginary parts of $\hat{V}$.

Characteristic impedance

So far we have discussed the propagation of waves on the line solely in terms of the voltage. The corresponding solution for the current can be obtained by substituting

$$V = V_1 \exp[\mathrm{j}(\omega t - kx)] \qquad \text{and} \qquad I = I_1 \exp[\mathrm{j}(\omega t - kx)]$$

into Equation (7.1), with the result

$$-\mathrm{j}kV_1 = -\mathrm{j}\omega L I_1$$

so that

$$I_1 = (k/\omega L)\, V_1$$

$$= V_1/Z_0 \tag{7.16}$$

where

$$Z_0 = \omega L/k = \sqrt{(L/C)} \tag{7.17}$$

This constant has the dimensions of resistance. It is known as the *characteristic impedance* of the line. Equation (6.54) shows that Z_0 depends upon the geometry of the line and the properties of the material between the conductors. For the lossless line considered here Z_0 is a real quantity, that is, a pure resistance, because the voltage and current are in phase with each other. It can be shown that if, on the other hand, the line is lossy, then the voltage and current are no longer in phase with each other and, consequently, Z_0 has an inductive or capacitive component. The physical significance of the characteristic impedance becomes clearer when we consider the reflection of waves from the termination of the line.

The characteristic impedance of a uniform line is the limit of the iterative impedance of a ladder network as the lengths of the section tend to zero. For a discussion of ladder networks consult Bleaney and Bleaney.*

Reflection of waves at the end of a line

Figure 7.3 shows a transmission line of characteristic impedance Z_0 which is terminated by an impedance Z at $x = 0$. The wave incident on the termination is

$$V = V_\mathrm{i} \exp[\mathrm{j}(\omega t - kx)] \tag{7.18}$$

We must assume, until it has been shown otherwise, that some of the incident wave will be reflected as the wave

$$V = V_\mathrm{r} \exp[\mathrm{j}(\omega t + kx)] \tag{7.19}$$

The amplitudes V_i and V_r are complex. The corresponding currents are

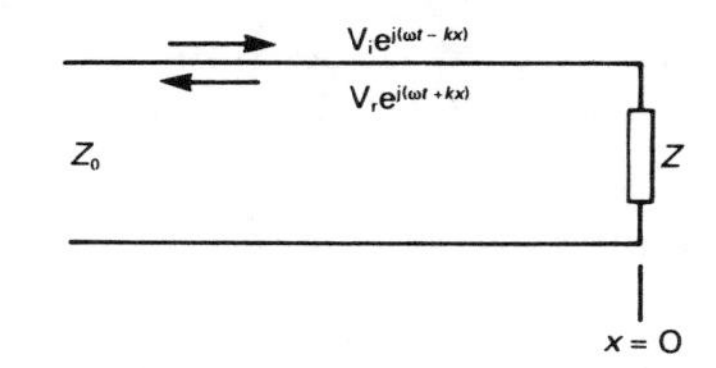

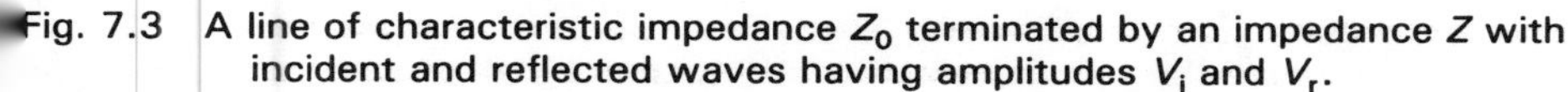
Fig. 7.3 A line of characteristic impedance Z_0 terminated by an impedance Z with incident and reflected waves having amplitudes V_i and V_r.

$$I = (V_i/Z_0) \exp[j(\omega t - kx)] \tag{7.20}$$

and

$$I = -(V_r/Z_0) \exp[j(\omega t + kx)] \tag{7.21}$$

as can be shown by substituting these expressions in Equation (7.1). We shall see later that the negative sign in Equation (7.21) indicates that the power in the reflected wave is travelling in the negative x direction. At $x = 0$ the voltage is

$$V = (V_i + V_r) \exp(j\omega t) \tag{7.22}$$

and the current is

$$I = ((V_i - V_r)/Z_0) \exp(j\omega t) \tag{7.23}$$

by adding together the voltages and currents for the incident and reflected waves. But the ratio of the voltage to the current at $x = 0$ must equal the termination impedance Z. Therefore

$$Z = \frac{V_i + V_r}{V_i - V_r} Z_0 \tag{7.24}$$

which can be rearranged to give

$$\frac{V_r}{V_i} = \frac{Z - Z_0}{Z + Z_0} = \Gamma \tag{7.25}$$

The voltage reflection coefficient of a load can be measured by a system known as a reflectometer. Devices known as directional couplers sample the incident and reflected power levels and their ratio is calculated and displayed by a special electronic instrument. See R.G. Carter.*

where Γ is known as the **voltage reflection coefficient**. Equation (7.25) shows that the amplitude of the reflected wave is zero when $Z = Z_0$. When this condition is satisfied the line is said to be **matched.**

When the line is terminated by a load equal to its characteristic impedance the instantaneous power absorbed in the resistor is

$$W = VI = (|V_i|^2/Z_0) \cos^2(\omega t) \tag{7.26}$$

The mean power is obtained by averaging Equation (7.26) over one complete cycle, giving

$$\overline{W} = |V_i|^2/2Z_0 \tag{7.27}$$

Since there is no reflected wave when $Z = Z_0$ it follows that Equation (7.27) is the mean power flow in the incident wave. If the load resistor were replaced by a semi-infinite section of transmission line having the same characteristic impedance there would still be no reflected wave in the part of the line to the left of $x = 0$. Thus the characteristic impedance is the input impedance of the semi-infinite line. Since we could just as well have replaced the load resistor by a section of line of finite length terminated by Z_0 without producing a reflected wave it follows that the input impedance of a transmission line terminated by its characteristic impedance is also Z_0.

When the line is not terminated by its characteristic impedance the voltage at a general point on it is

$$V = V_i \exp[j(\omega t - kx)] + \Gamma V_i \exp[j(\omega t + kx)] \tag{7.28}$$

The amplitude of the voltage which would be measured at this point is

$$|V| = |V_i|\,|\exp(-jkx)|\,|1 + \Gamma\exp(2jkx)|$$
$$= |V_i|\,|1 + \Gamma\exp(2jkx)| \qquad (7.29)$$

since $|\exp(-jkx)| = 1$. Equation (7.29) shows that the amplitude of the voltage varies along the line between the maximum and minimum values

$$V_{max} = |V_i|(1 + |\Gamma|) \qquad (7.30)$$

$$V_{min} = |V_i|(1 - |\Gamma|) \qquad (7.31)$$

The ratio of these two is easily measured with a sliding probe and a suitable detector. It is known as the **voltage standing wave ratio** (VSWR), and its value is given by

$$S = \frac{1 + |\Gamma|}{1 - |\Gamma|} \qquad (7.32)$$

The VSWR on a line can be measured by using a slotted section of line which allows the electric field strength to be sampled by a probe. The signal from the probe is displayed on an instrument known as a **VSWR meter**. See R.G. Carter.*

The probe, therefore, detects a signal whose amplitude varies periodically with position along the line. The wavelength of the standing wave observed is given by $2kx = 2\pi$ from Equation (7.29), that is

$$x = \lambda/2 \qquad (7.33)$$

where λ is the wavelength of the travelling waves.

Pulses on transmission lines

So far we have assumed that the signals propagating on the transmission lines are sinusoidal. It is, of course, possible to consider non-sinusoidal signals by using Fourier synthesis to construct them from sine waves of different frequencies. But pulses, which are particularly important now that digital data transmission is being used more and more, can most easily be studied directly.

For a fuller treatment of this subject consult Carter, G.W. and Richardson, A., *Techniques of Circuit Analysis*, Cambridge University Press (1972).

It is easy to show by substitution that any functions having the form

$$V = V_0 f(x - v_p t) \qquad \text{and} \qquad I = I_0 f(x - v_p t) \qquad (7.34)$$

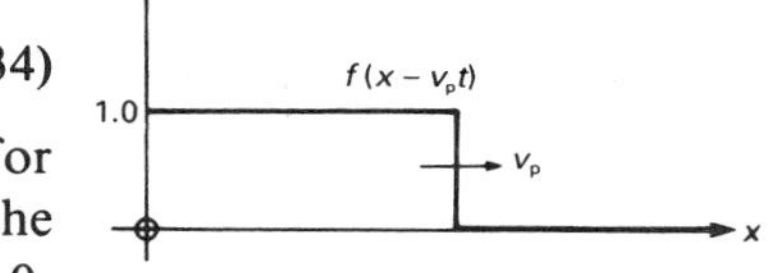

are solutions of Equations (7.1)–(7.4). The function f can have any form, but for the study of pulses it is taken to be the step function shown in the figure in the margin. This function is defined to be 0 for $(x - v_p t) > 0$ and 1 for $(x - v_p t) \leqslant 0$. Substituting the voltage and current given in Equation (7.34) into Equation (7.1) gives

$$V_0 f' = v_p L I_0 f'$$

where f' is the derivative of f, so that

$$V_0 = Z_0 I_0 \qquad (7.35)$$

as before, from Equations (7.11) and (7.17).

Figure 7.4(a) shows a voltage source with internal impedance Z_s connected to a semi-infinite transmission line of characteristic impedance Z_0. If the line is uncharged until the switch is closed at $t = 0$, then a wave having the form shown in Fig. 7.4(b) will propagate down the line with velocity v_p. As v_p is independent of frequency the step propagates without any change in its shape. A line for which this is true is said to be **non-dispersive**. Conversely, if v_p depends on frequency,

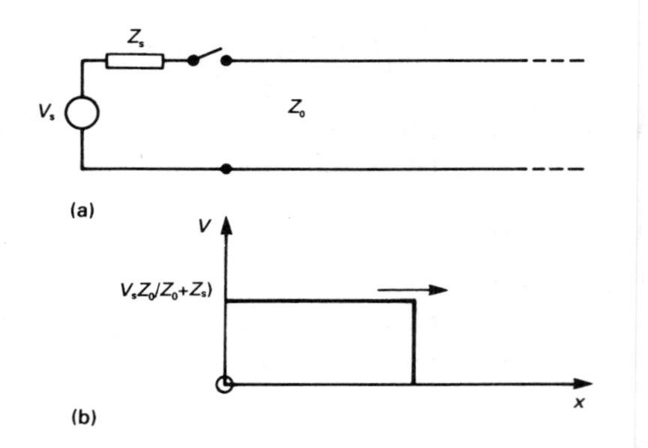

Fig. 7.4 A semi-finite transmission line of characteristic impedance Z_0 is connected to a source of impedance Z_s by a switch at $t = 0$. The result is a step voltage wave which travels down the transmission line as shown.

the line is **dispersive,** the different Fourier components of the step travel at different speeds, and the shape of the step changes as it propagates. The voltage and current at the start of the line must satisfy Equation (7.35), so that

$$V_0 = \frac{Z_0}{Z_0 + Z_s} V_s \qquad (7.36)$$

by the potential-divider rule.

Reflection of pulses at the end of a line

When a transmission line is terminated by a resistance R, part of an incident pulse is normally reflected. The voltage and current at any point on the line are given by

$$V = V_i + V_r$$

and

$$I = I_i - I_r = (V_i - V_r)/Z_0$$

To satisfy the boundary conditions at the termination

$$R = \frac{V}{I} = \frac{V_i + V_r}{V_i - V_r} Z_0$$

so that

$$\frac{V_r}{V_i} = \frac{R - Z_0}{R + Z_0} \qquad (7.37)$$

From 7.37 it is clear that there are three possible conditions:

1. If $R > Z_0$, V_r is positive so that the reflected wave is added to the incident wave.
2. If $R < Z_0$, V_r is negative and the reflected wave is subtracted from the incident wave.

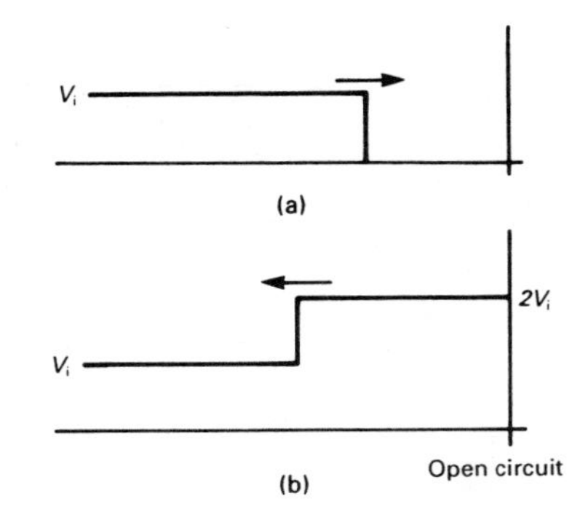

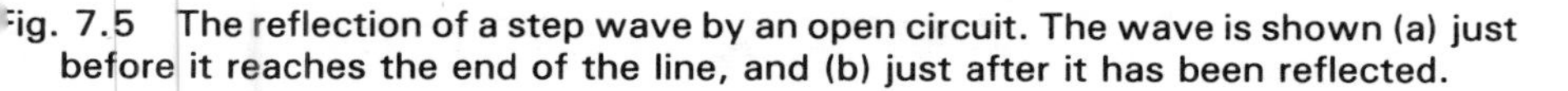
Fig. 7.5 The reflection of a step wave by an open circuit. The wave is shown (a) just before it reaches the end of the line, and (b) just after it has been reflected.

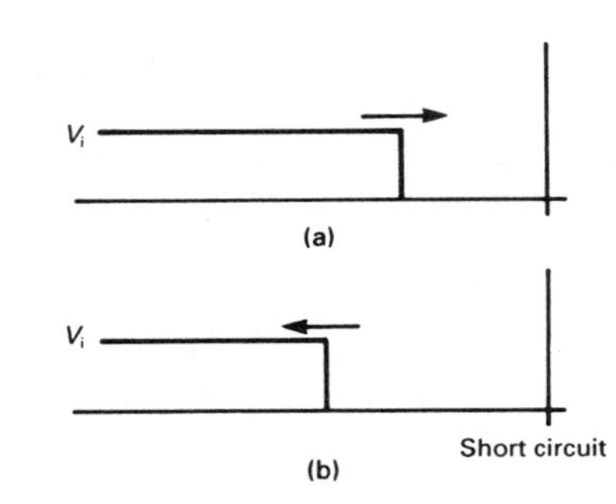

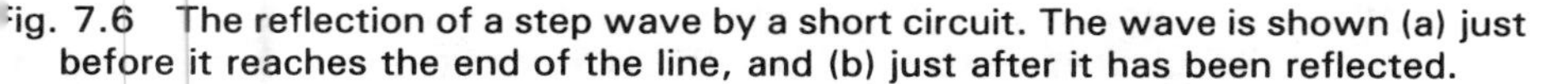
Fig. 7.6 The reflection of a step wave by a short circuit. The wave is shown (a) just before it reaches the end of the line, and (b) just after it has been reflected.

3. If $R = Z_0$, V_r is zero and the incident wave is completely absorbed by the termination.

Conditions (1) and (2) can be illustrated by the extreme cases of termination of a line by an open circuit and a short circuit. Figure 7.5 shows what happens when here is an open circuit. Figure 7.5(a) shows the situation just before the incident wave has reached the open circuit. When it reaches it a reflected wave is generated o satisfy the boundary conditions. The amplitude of the reflected wave is found o be equal to that of the incident wave from Equation (7.37). A short time later he situation is as shown in Fig. 7.5(b), with the reflected wave superimposed on he incident wave. In the same way Fig. 7.6(a) shows a wave approaching a short ircuit. This time the wave is inverted on reflection so that the sum of the wave amplitudes is always zero at the short circuit. A short time after the reflection of he wave the situation is as shown in Fig. 7.6(b).

It is instructive to consider the effects of these reflections at the input of the ransmission line. To make things simpler we will assume that the source is natched to the line so that $Z_s = Z_0$ as shown in Fig. 7.7(a). Figure 7.7(b) shows how the voltage at the start of the line varies with time when the termination is an open circuit. Initially $V_A = \frac{1}{2} V_s$ as the source feeds current into the line to set up the incident wave. This wave travels down the line in time T. At time $2T$ the eflected wave returns to A so that $V_A = V_s$. The reflected wave is completely

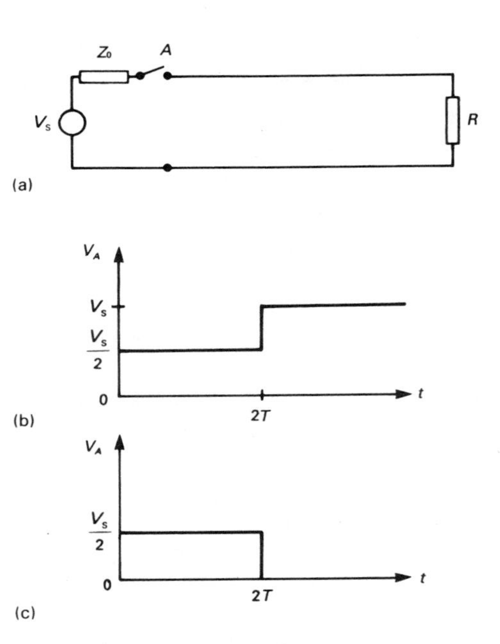

Fig. 7.7 Variation of the voltage at the input of a transmission line with time. Th line has characteristic impedance Z_0, is connected to a matched source and to a mis matched load. When the termination is an open circuit the voltage varies as show in (b). (c) shows what happens when Z is a short circuit.

absorbed by the source impedance, so no further change takes place and the lin is charged to the full source voltage. If the line is terminated by a short circuit th voltage varies with time, as shown in Fig. 7.7(c). There is an initial period whe the voltage at A is not zero, but this is cancelled at time $2T$ by the return of th reflected wave. According to elementary circuit theory the voltages would tak their final values as soon as the switch was closed. The true situation can b described by saying that, until the return of the reflected wave, the source has n information about the magnitude of the terminating impedance. The step canno travel faster than the speed of light, so it always takes a finite time for the syster to reach a steady state.

It is shown in Worked Example 7.3 that the phase velocity of waves on a polythene-insulated coaxial cable is about two-thirds of the velocity of light.

The transmission of a short pulse along a line can be investigated by considerin the superposition of positive and negative step functions separated by a shor interval of time.

Worked Example 7.1

A transmission line of characteristic impedance Z_0 is connected to a matche source for a short time τ by a switch to produce a pulse on the line. If the propaga tion time along the line is T and the line is terminated by a resistance $R = \frac{1}{2} Z_($ investigate the variation of voltage with time at each end of the line.

Solution The circuit is shown in Fig. 7.8(a). At the start of the line there is voltage pulse of amplitude $\frac{1}{2} V_s$, as shown in Fig. 7.7(b). After time T this puls reaches the other end of the line where the reflection coefficient is $-\frac{1}{3}$ fror Equation (7.37) so the reflected pulse has an amplitude of $-\frac{1}{6} V_s$. The puls

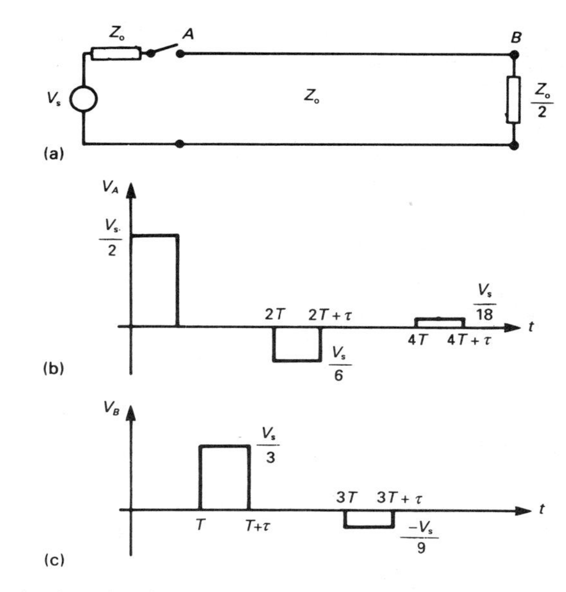

Fig. 7.8 A matched source is connected to a transmission line of impedance Z_0 for a short time τ. (b) and (c) show how multiple reflections of the pulse occur at the open circuit at the source and at the mismatched load. The sequence of events is discussed in Worked Example 7.1.

which appears across the load therefore has amplitude $\frac{1}{3}V_s$, as shown in Fig. 7.8(c). After time $2T$ the reflected pulse returns to A, which is now an open circuit, so it is completely reflected to return to B at time $3T$, and so on. The result of closing the switch for a short time is to produce not just one pulse at B but a whole string of pulses with gradually decreasing amplitudes. If the line is short so that $T \ll \tau$, then this effect shows itself as a blurring of the edges of the pulse. Evidently care has to be taken with the matching of lines for the digital transmission of data if errors in the information received are not to occur.

Worked Example 7.2

A voltage step generator whose source impedance is 50 Ω is connected to an oscilloscope by 1 m of 50 Ω polythene-insulated coaxial cable. If the input impedance of the oscilloscope is 1 MΩ in parallel with 10 pF, find the waveforms appearing at the input of the oscilloscope and at the source.

Solution The relative permittivity of polythene is 2.25 so the phase velocity of the line is $0.2 \times 10^9\ \mathrm{m\,s^{-1}}$ and the transit time for signals is 5 ns. Initially the source sees the cable as a matched load so the incident wave amplitude $V_+ = 0.5V_s$. When this pulse reaches the oscilloscope the instantaneous boundary conditions require that the voltage should be zero. When the system eventually reaches equilibrium the oscilloscope input voltage will be equal to V_s because the input resistance of the oscilloscope is much greater than the characteristic impedance of the cable. This voltage is the sum of the incident and reflected waves so the reflected wave must rise from $V_- = -0.5\ V_s$ at $t = 5$ ns to $+0.5\ V_s$ in the steady state. The current and voltage at the input of the oscilloscope are related to each other by

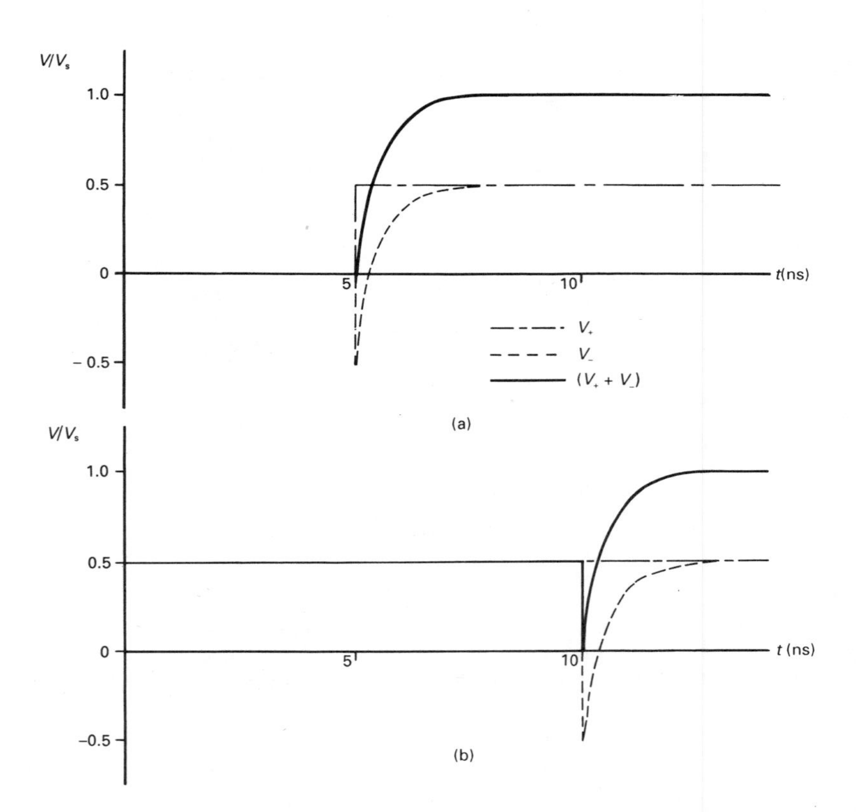

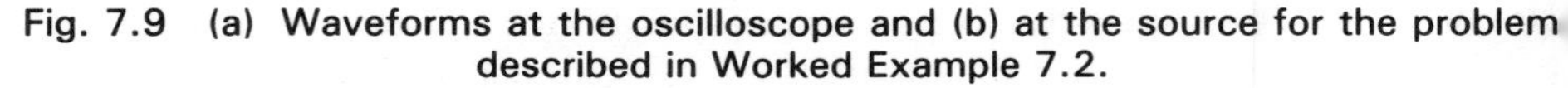

Fig. 7.9 (a) Waveforms at the oscilloscope and (b) at the source for the problem described in Worked Example 7.2.

$$I_+ - I_- = C\frac{\mathrm{d}}{\mathrm{d}t}(V_+ + V_-)$$

where C is the input capacitance of the oscilloscope so that

$$V_+ - V_- = Z_0 C\frac{\mathrm{d}V_-}{\mathrm{d}t}$$

since $\frac{\mathrm{d}V_+}{\mathrm{d}t} = 0$. Because Z_0 is much less than the input resistance of the oscilloscope the time constant is $Z_0C = 0.5$ ns. The reflected wave is completely absorbed by the source because the source is matched to the cable. Thus the waveforms at the oscilloscope input and at the source are as shown in Fig. 7.9.

The response of the system to short pulses could be found by superimposing positive and negative steps at appropriate time intervals.

Transformation of impedance along a transmission line

Consider a transmission line having characteristic impedance Z_0 which is terminated at $x = 0$ by an impedance Z_L. At the point on the line which is l from the load, $x = -l$ and the voltage is given by

$$V = V_i \exp[j(\omega t + kl)] + \Gamma V_i \exp[j(\omega t - kl)] \tag{7.38}$$

The corresponding expression for the current is

$$I = (V_i/Z_0)\exp[j(\omega t + kl)] - \Gamma(V_i/Z_0)\exp[j(\omega t - kl)] \tag{7.39}$$

where

$$\Gamma = (Z_L - Z_0)/(Z_L + Z_0) \tag{7.40}$$

from Equation (7.25). The apparent impedance at this point on the line is given by

$$\begin{aligned} Z_L' &= V/I \\ &= \frac{\exp(jkl) + \Gamma\exp(-jkl)}{\exp(jkl) - \Gamma\exp(-jkl)} Z_0 \end{aligned} \tag{7.41}$$

After substituting for Γ and rearranging, we get

$$\begin{aligned} \frac{Z_L'}{Z_0} &= \frac{Z_L[\exp(jkl) + \exp(-jkl)] + Z_0[\exp(jkl) - \exp(-jkl)]}{Z_L[\exp(jkl) - \exp(-jkl)] + Z_0[\exp(jkl) + \exp(-jkl)]} \\ &= \frac{Z_L\cos(kl) + jZ_0\sin(kl)}{jZ_L\sin(kl) + Z_0\cos(kl)} \\ &= \frac{Z_L + jZ_0\tan(kl)}{jZ_L\tan(kl) + Z_0} \end{aligned} \tag{7.42}$$

Calculations using Equation (7.42) are made much easier by the use of the Smith chart. There is not room to discuss this here, but further information can be found in R.G. Carter.*

This is a very important result. It shows that if the length l of the transmission line is of the same order of magnitude as the wavelength $2\pi/k$, or greater, then the apparent impedance at the input of the line is not normally equal to the load impedance. The usual assumption of circuit theory is that the impedances are independent of the lengths of the connecting wires. We now see that this is just the limit of Equation (7.42) when kl tends to zero. The significance of Equation (7.42) can be explained in a slightly different way by considering Fig. 7.10. This shows two transmission lines, one of which is longer than the other by l. The longer line is terminated by a load Z_L and the shorter by Z_L', where Z_L' is given by Equation (7.42). At the plane A–A the impedance presented to the incident wave is the same

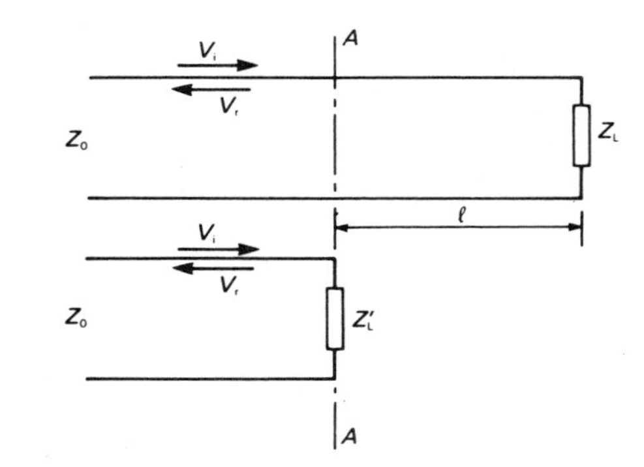

Fig. 7.10 The apparent terminating impedance of a transmission line varies with position along the line. If two lines are terminated by Z_L and Z_L' as calculated from Equation (7.42), then the amplitude and phase of the reflected wave to the left of the plane A–A is the same on both lines.

on both lines. Thus the reflected wave to the left of A–A must have the same amplitude and phase on both lines. It follows that Z_L and Z_L' must reflect the incident wave with the same amplitude but with a phase difference of $\exp(2jkl)$, since that is the phase change from A–A to Z_L and back.

Worked Example 7.3 A transmission line has a characteristic impedance of 50 Ω, and at 1 GHz the wavelength of the signal on the line is 150 mm. Given that the line is 20 mm long and the load impedance is 100 Ω, find the impedance at the input of the line.

Solution $kl = 2\pi \times 20/150$, so $\tan(kl) = 1.11$.
From Equation (7.42) the input impedance is

$$\begin{aligned} Z_L' &= \frac{100 + 1.11 \times 50j}{1.11 \times 100j + 50} \times 50 = \frac{100 + 55.5j}{2.22j + 1} \\ &= (114.4\,e^{0.507j})/(2.435\,e^{1.148j}) \\ &= 47.0\,e^{-0.641j} = 37.7 - 28.1j\ \Omega \end{aligned}$$

The quarter-wave transformer

A case of considerable practical importance is obtained when the line is exactly one quarter of a wavelength long. Setting $kl = \pi/2$ in Equation (7.42) gives

$$Z_L'/Z_0 = Z_0/Z_L$$

or

$$Z_0^2 = Z_L/Z_L'. \tag{7.43}$$

This result allows us to match a load to a line by inserting a short section of a line having a different impedance between them, as shown in Fig. 7.11. The short length of line must be a quarter wavelength long, and it must have the characteristic impedance given by (7.43). The same technique can be used to ensure a match between two transmission lines which have different impedances. The quarter-wave transformer suffers from the difficulty that it works exactly only at a single frequency. This problem can be overcome by cascading a number of quarter-wave transformers to produce a broad-band match.

Reference tables for the design of broadband quarter-wave transformers were published by Young, *Tables for cascaded homogeneous quarter wave transformers, IRE Trans.* **MTT-7**, 233–7 (1959) and **MTT-8**, 243–4 (1960).

Two special cases arise when the terminating impedance of the transformer is either an open circuit or a short circuit. If there is an open circuit, then $Z_L = \infty$ and (7.42) becomes

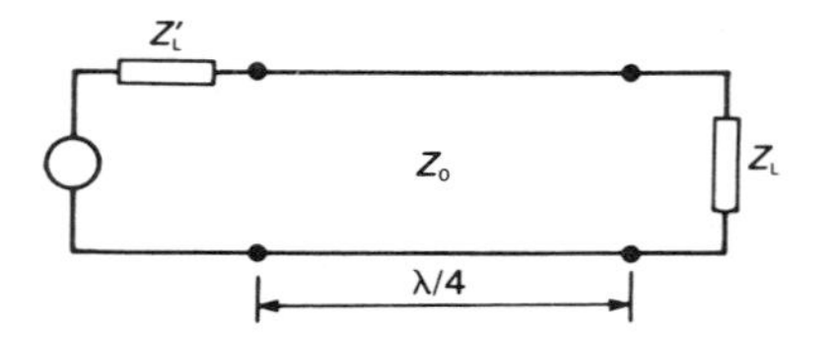

Fig. 7.11 A source of impedance Z_L' can be matched to a load of impedance Z_L by using a quarter wavelength of a transmission line whose characteristic impedance is $Z_0 = \sqrt{(Z_L Z_L')}$ as a transformer.

$$Z_L'/Z_0 = 1/(j \tan kl)$$

so that Z_L' tends to zero as kl tends to $\pi/2$. Thus a quarter-wave transformer transforms an open circuit into a short circuit. The converse is also true. If the termination of the transformer is a short circuit, then the impedance at its input tends to infinity. In either case there is a total reflection of the incident power.

Field description of transmission lines

The simplest practical transmission line from the field point of view is the coaxial cable. We shall therefore take it as the example to be discussed, but the methods applied, and the conclusions reached, are applicable to all types of two-wire line. A typical cable has a solid copper inner conductor and a braided outer conductor separated by a uniform cylindrical layer of polythene. To make the problem tractable we assume that the braided outer conductor can be represented with sufficient accuracy by a continuous conducting cylinder. Taking the radii of the inner and outer conductors as a and b, respectively, we recall that the capacitance per unit length is

$$C = 2\pi\epsilon/\ln(b/a) \tag{7.44}$$

a result which can be deduced from the result of Worked Example 2.3. The inductance per unit length is

$$L = (\mu_0/2\pi)\ln(b/a) \tag{7.45}$$

from Worked Example 6.2.

Substituting these expressions for C and L into Equations (7.11) and (7.17) we get

$$v_p = 1/\surd(\epsilon\mu_0) \tag{7.46}$$

and

$$Z_0 = \frac{1}{2\pi}\sqrt{\frac{\mu_0}{\epsilon}}\ln(b/a) \tag{7.47}$$

A very interesting result is obtained for the special case of an air-spaced line. In that case

$$\begin{aligned} v_p &= 1/\surd(\epsilon_0\mu_0) \\ &= (8.854 \times 10^{-12} \times 4\pi \times 10^{-7})^{-\frac{1}{2}} \\ &= 2.998 \times 10^8 \text{ m s}^{-1} \end{aligned}$$

This throws light on the *LCRZ* analogy discussed in Chapter 6.

which is the experimental value for the velocity of light in free space. This suggests the existence of a link between electromagnetism and optics, and the possibility that light is an electromagnetic phenomenon. Since the relative permittivity of any medium is greater than unity, it follows that the phase velocity of the waves on a transmission line is always less than or equal to the velocity of light.

Worked Example 7.4

A polythene-insulated coaxial cable has a characteristic impedance of 50 Ω, and an inner conductor 1 mm in diameter. Calculate the diameter of the outer conductor and the wavelength on the line at a frequency of 1 GHz.

Solution The relative permittivity of polythene is 2.25. Hence, from Equation (7.47),

$$\ln(b/a) = 50 \times 2\pi \times (2.25 \times 8.854 \times 10^{-12}/4\pi \times 10^{-7})^{\frac{1}{2}} = 1.25$$

so the diameter of the outer conductor should be $1.0 \times e^{1.25} = 3.5$ mm.

The phase velocity is

$$(2.25 \times 8.854 \times 10^{-12} \times 4\pi \times 10^{-7})^{-\frac{1}{2}} = 2.0 \times 10^{8}\ \mathrm{m\ s^{-1}}$$

so, at 1 GHz the wavelength is given by $\lambda = v_p/f = 0.2$ m.

The electric and magnetic fields in a coaxial line

To find expressions for the electric and magnetic fields within a coaxial line which is carrying a wave along it we apply Gauss' theorem and the magnetic circuit law. This approach involves us in making some assumptions about the fields in the line. Taking the electric field first, we assume that, as in the electrostatic case, only the radial component is present. The charge per unit length on the centre conductor is given by

$$q = C V_0 \exp[\mathrm{j}(\omega t - kz)] \tag{7.48}$$

so, applying Gauss' theorem

$$\begin{aligned} E_r &= \frac{q}{2\pi\epsilon r} \\ &= \frac{C V_0}{2\pi\epsilon} \frac{1}{r} \exp[\mathrm{j}(\omega t - kz)] \end{aligned}$$

Substituting for the capacitance per unit length from Equation (7.44) gives

$$E_r = \frac{V_0}{\ln(b/a)} \frac{1}{r} \exp[\mathrm{j}(\omega t - kz)] \tag{7.49}$$

The magnetic field is given by

$$\begin{aligned} H_\theta &= I/2\pi r \\ &= \frac{1}{2\pi r} \frac{V_0}{Z_0} \exp[\mathrm{j}(\omega t - kz)] \end{aligned}$$

Substituting for Z_0 from Equation (7.47)

$$\begin{aligned} H_\theta &= \sqrt{\frac{\epsilon}{\mu_0}} \frac{V_0}{\ln(b/a)} \frac{1}{r} \exp[\mathrm{j}(\omega t - kz)] \\ &= \sqrt{\frac{\epsilon}{\mu_0}}\, E_r \end{aligned} \tag{7.50}$$

These fields are shown in Fig. 7.12. We note that the ratio of the strengths of the electric and magnetic fields is

$$E_r/H_\theta = \sqrt{\frac{\mu_0}{\epsilon}} = Z_w \tag{7.51}$$

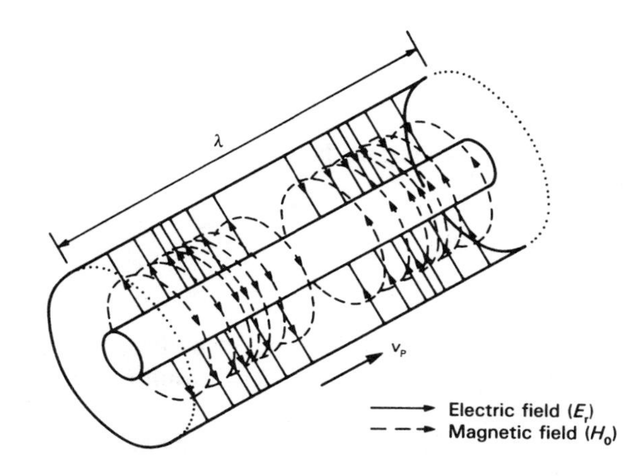

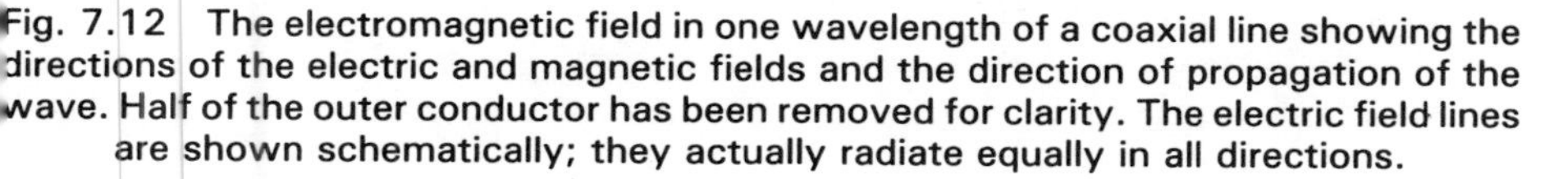

Fig. 7.12 The electromagnetic field in one wavelength of a coaxial line showing the directions of the electric and magnetic fields and the direction of propagation of the wave. Half of the outer conductor has been removed for clarity. The electric field lines are shown schematically; they actually radiate equally in all directions.

This quantity, which depends only upon the properties of the material filling the line, has the dimensions of resistance. It is known as the **wave impedance** of the wave.

Although these results have been derived by making assumptions about the directions of the fields, they are confirmed by more rigorous analysis. It is found that, for any two-wire transmission line, the electric and magnetic field vectors are perpendicular to each other and to the direction of propagation of the wave. Such waves are known as transverse electric and magnetic (TEM) waves.

It can be shown that the distribution of electric and magnetic fields in a TEM wave on any two-wire line can be found by assuming d.c. conditions. See R.G. Carter.* This is the theoretical justification of the *LCRZ* analogy discussed in Chapter 6.

Power flow in a coaxial line

When discussing the storage of energy in capacitors and inductors we saw that it is sometimes useful to think of the energy as being distributed throughout the electric or magnetic fields. In the same way we can think of the power flowing down a coaxial cable as being distributed throughout the field. At any point the electric energy density is

$$w_{\mathrm{E}} = \tfrac{1}{2}\boldsymbol{D}\cdot\boldsymbol{E} = \tfrac{1}{2}\epsilon E_{\mathrm{r}}^2 \tag{7.52}$$

while the magnetic energy density is

$$w_{\mathrm{M}} = \tfrac{1}{2}\boldsymbol{B}\cdot\boldsymbol{H} = \tfrac{1}{2}\mu_0 H_\theta^2 \tag{7.53}$$

Making use of Equation (7.51) we can write the total energy density as

$$\begin{aligned} w &= w_{\mathrm{E}} + w_{\mathrm{M}} \\ &= \tfrac{1}{2}\epsilon E_{\mathrm{r}}^2 + \tfrac{1}{2}\mu_0\left(\frac{\epsilon}{\mu_0}E_r^2\right) \\ &= \epsilon E_{\mathrm{r}}^2 \end{aligned} \tag{7.54}$$

Now at any point within the line the fields vary sinusoidally with time. Therefore the time average of the stored energy is

$$\overline{w} = \tfrac{1}{2}\epsilon E_r^2$$
$$= \frac{\epsilon V_0^2}{2(\ln(b/a))^2}\frac{1}{r^2} \tag{7.55}$$

The whole field pattern is moving in the positive z direction with velocity v_p, so the power density is

$$P = \overline{w}v_p \tag{7.56}$$
$$= \tfrac{1}{2}E_r(\epsilon v_p E_r)$$
$$= \tfrac{1}{2}E_r H_\theta \tag{7.57}$$

making use of Equations (7.46) and (7.51). We shall see in Chapter 8 that this is an example of the application of a more general result relating to the power flow in electromagnetic waves.

To check the equivalence of the field and circuit descriptions of the problem we can calculate the total power flow by integrating Equation (7.56) over the space between the conductors. Then

$$P_{\text{tot}} = \int_a^b \frac{\epsilon v_p V_0^2}{2(\ln(b/a))^2}\frac{2\pi r}{r^2}\,\mathrm{d}r$$
$$= \tfrac{1}{2}V_0^2\sqrt{\frac{\epsilon}{\mu_0}}\frac{2\pi}{\ln(b/a)} = \tfrac{1}{2}V_0^2/Z_0 \tag{7.58}$$

from the expression for Z_0 given in Equation (7.47), so the power is identical to that calculated by circuit methods (Equation (7.27)).

Summary

In this chapter transmission lines have been introduced through a discussion of distributed circuits. The voltages and currents were found to be governed by the wave equation, and it was demonstrated that solutions could be found which could be interpreted as travelling waves. The power carried by these waves was expressed in terms of the voltage on the line and its characteristic impedance. It was demonstrated that, unless a line is matched by being terminated by a load equal to its characteristic impedance, some of the power in a wave incident upon the termination is reflected back down the line. This reflection produces a partial standing wave on the line. The effect of the reflected wave is to make the input impedance of a line depend upon its length as well as upon the terminating impedance. Thus it was shown that, at high frequencies when the lengths of lines are comparable with the wavelengths of the signals on them, it is no longer possible to treat circuits in terms of lumped components connected together by wires of arbitrary length. The transformation of impedance by a short length of a transmission line can be put to use in the form of a quarter-wave transformer to match a load to a line.

It is also important for lines to be correctly matched when the signals are in the form of short pulses. Incorrect matching results in reflections of the pulses. If a line is not matched at either end, multiple reflections can occur so that a whole train of pulses is produced by transmitting a single pulse down the line.

Transmission lines were also considered from the point of view of the electric and magnetic fields within them. It was shown that the phase velocity and characteristic impedance of a line at a given frequency can be calculated from field considerations.

Expressions were found for the electric and magnetic fields within a coaxial line, and it was demonstrated that the flow of power down the line could be regarded as that of an electromagnetic wave propagating in the space between the conductors. It was found that, for an air-spaced line, these waves propagate with a phase velocity equal to the speed of light. The equivalence of the circuit and field approaches to the description of waves on transmission lines was demonstrated by showing that both give the same expression for the power flow.

Problems

7.1 A transmission line is terminated by a resistive load such that the VSWR on the line is 2.0. Calculate the voltage reflection coefficient of the load, the ratio of the load impedance to the characteristic impedance of the line, and the ratio of the reflected signal to the incident signal, expressing your answer in decibels.

7.2 Sketch graphs of the variation of voltage with time at A and B in the circuit of Fig. 7.8 if the load resistor is replaced by one having an impedance $2Z_0$.

7.3 A solenoid is wound uniformly at 318 turns per metre on an insulating cylindrical former 100 mm in diameter and 2.5 m long. Calculate its inductance per unit length. Ten tapping points are made on the solenoid at regular intervals, and each is connected to earth through a 0.001 μF capacitor. The resulting network is a cascade of symmetrical tee sections as shown in Fig. 7.13. This line is used as the pulse-forming network for a high-power radar transmitter in the following manner. The line is charged to a potential difference of 20 kV with both ends open-circuited; a matched load is then connected across one end and the line is discharged through it. Assuming that the line can be treated as a uniform transmission line, calculate the amplitude and duration of the voltage pulse supplied to the load. Calculate also the current flowing in the load during the pulse, and the total pulse energy.

In a radar transmitter the load is a transmitter tube (a magnetron, klystron or travelling wave tube). Because the pulse-forming network is not a continuous line, a series of ripples appears on top of the pulse.

7.4 A coaxial cable with a characteristic impedance of 50 Ω is insulated with a dielectric of relative permittivity 2.7. Calculate the wavelength of signals on the line at 500 MHz, 1 GHz and 2 GHz. Given that a 1 metre length of this cable is terminated in a 75 Ω resistor, calculate the input impedance of the cable at each of the three frequencies given above.

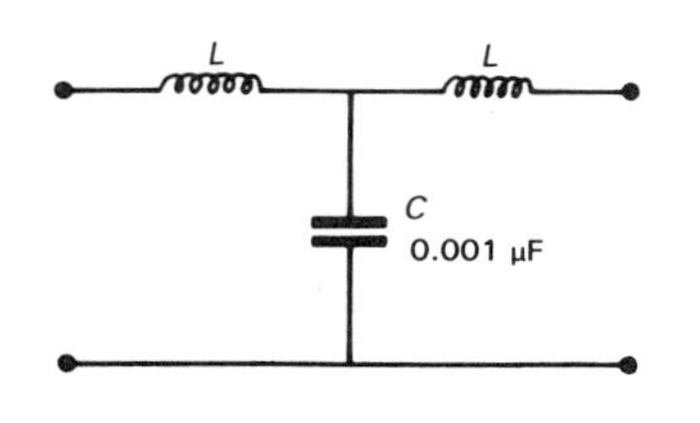

Fig. 7.13

7.5 A 50 Ω air-spaced coaxial cable has an outer conductor with 10 mm inside diameter. Calculate the diameter of the inner conductor. A quarter-wave transformer is needed to match this cable to a 75 Ω load at 1 GHz. Assuming that the outer diameter of the transformer is the same as that of the cable, calculate the dimensions of the transformer.

7.6 Show that an air-spaced coaxial cable is correctly matched if a sheet of resistive card having a surface resistance of 377 Ω per square is placed across its end.

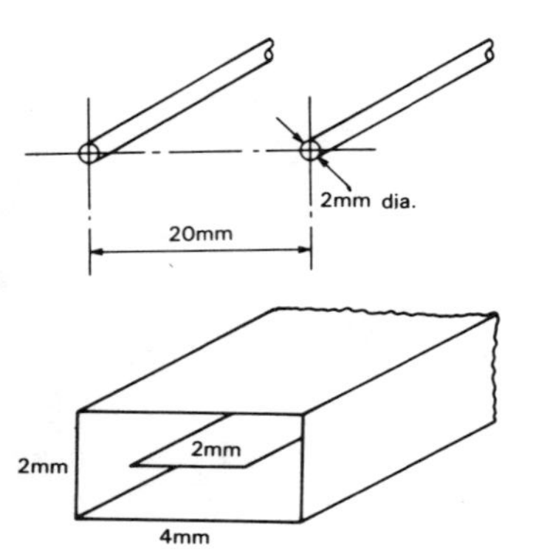

7.7 Find the phase velocity and characteristic impedance for the air-spaced two-wire transmission line shown in the top figure in the margin.

7.8 Find the phase velocity and characteristic impedance for the square coaxial transmission line shown in the bottom figure in the margin if the space between the conductors is filled with a material having a relative permittivity of 4.0.

Maxwell's equations and electromagnetic waves

8

Objectives

- ☐ To show how Maxwell removed an inconsistency in the magnetic circuit law by introducing the idea of the displacement current.
- ☐ To derive the differential forms of the magnetic circuit law and Faraday's law, and to introduce the curl of a vector.
- ☐ To present Maxwell's equations in both integral and differential form.
- ☐ To demonstrate how the existence of plane electromagnetic waves in free space can be deduced from Maxwell's equations.
- ☐ To show that the power flow in an electromagnetic field can be calculated by integrating the Poynting vector over a closed surface.

In Chapter 7 we saw how it is possible for electromagnetic waves to propagate along a transmission line such as a coaxial cable. We shall now proceed to show that electromagnetic waves can exist independently of any system of conductors and travel through free space. This conclusion was first published by James Clerk Maxwell in 1873. The set of equations which bears his name forms a summary of all the topics treated in this book and demonstrates the symmetry and unity of electromagnetic theory. At the same time they are the foundation upon which modern electromagnetic theory is built. We are now so accustomed to the presence of radio, television and radar in the world that it is hard to realize how great Maxwell's contribution was to the understanding of the theory of electromagnetism. In his own day his theory was regarded as quite outrageous by many until its validity was demonstrated by Hertz's experiments some fifteen years later.

Maxwell's *Treatise on Electricity and Magnetism*, published in 1873, presented the first unified account of electromagnetic phenomena.

Maxwell's form of the magnetic circuit law

Maxwell noticed that, under certain circumstances, the magnetic circuit law in the form derived in Chapter 4 can give inconsistent results. This is readily demonstrated by considering the circuit shown in the figure in the margin – a charged capacitor discharging though a resistor. To apply the magnetic circuit law we must define a closed path C which encircles the wire and an open surface S which spans that path as a soap film can span a loop of wire. The figure in the margin attempts to show what is meant. If S is chosen so that it passes through the wire, then no problem arises and the magnetic field satisfies Equation (5.7), which is repeated here for convenience:

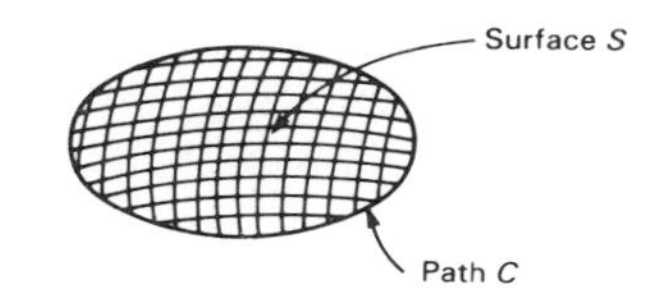

$$\oint_C \boldsymbol{H} \cdot \mathrm{d}\boldsymbol{l} = \iint_S \boldsymbol{J} \cdot \mathrm{d}\boldsymbol{A} \tag{8.1}$$

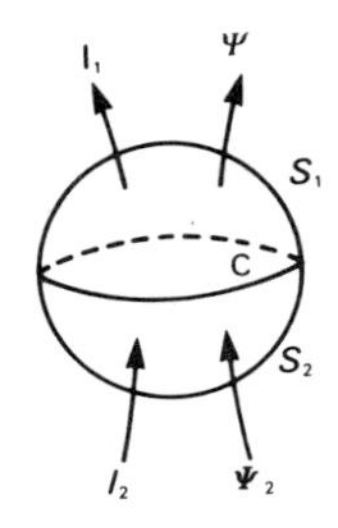

Fig. 8.1 Maxwell showed that the conduction current (I) passing through a closed loop (C) can be dependent upon the choice of the surface (S) over which it is calculated. The sum of the conduction and displacement currents ($I + \mathrm{d}\Psi/\mathrm{d}t$) is independent of the choice of surface.

But, if the surface S is chosen so that it passes between the plates of the capacitor, then $\boldsymbol{J} = \mathbf{0}$, implying that the line integral of $\boldsymbol{H}$ around C is also zero. This sounds nonsense! We expect the magnetic field around the circuit to be independent of the details of the means by which it has been calculated. To resolve this difficulty we consider two different surfaces S_1 and S_2 which both span C, as shown in Fig. 8.1, and which together form a closed surface. At a particular instant let the total charge enclosed within this surface be Q, and let currents I_1 and I_2 cross the two surfaces as shown. Now charge must be conserved, so the rate of change with time of the charge enclosed must be equal to the net current flow into the volume enclosed. That is,

$$I_2 - I_1 = \frac{\mathrm{d}Q}{\mathrm{d}t} \tag{8.2}$$

If we define Ψ_1 to be the flux of D through S_1 and Ψ_2 that through S_2, as shown in Fig. 8.1, then the application of Gauss' theorem gives

$$\Psi_1 - \Psi_2 = Q \tag{8.3}$$

Differentiating Equation (8.3) with respect to time and substituting for $\mathrm{d}Q/\mathrm{d}t$ in Equation (8.2) gives

$$I_1 + \frac{\mathrm{d}\Psi_1}{\mathrm{d}t} = I_2 + \frac{\mathrm{d}\Psi_2}{\mathrm{d}t} \tag{8.4}$$

This result can also be obtained by using Gauss' theorem to eliminate ρ from the continuity equation (3.13):

$$\oiint \boldsymbol{J}\cdot\mathbf{d}\boldsymbol{A} = -\frac{\partial}{\partial t}\iiint \rho\,\mathrm{d}v$$

$$= -\frac{\partial}{\partial t}\oiint \boldsymbol{D}\cdot\mathbf{d}\boldsymbol{A}$$

so that

$$\oiint \left(\boldsymbol{J} + \frac{\partial \boldsymbol{D}}{\partial t}\right)\cdot\mathbf{d}\boldsymbol{A} = 0$$

whence the result given in Equation (8.4).

This equation shows that the quantity ($I + \mathrm{d}\Psi/\mathrm{d}t$) is independent of the choice of the position of the surface S even when I is not. This quantity is known as the **total current**, and $\mathrm{d}\Psi/\mathrm{d}t$ is known as the **displacement current**. When the current density on the right-hand side of Equation (8.1) is replaced by the total current density ($\boldsymbol{J} + \partial\boldsymbol{D}/\partial t$), there results the equation

$$\oint_C \boldsymbol{H}\cdot\mathbf{d}\boldsymbol{l} = \iint_S \left(\boldsymbol{J} + \frac{\partial \boldsymbol{D}}{\partial t}\right)\cdot\mathbf{d}\boldsymbol{A} \tag{8.5}$$

This form of the magnetic circuit law is universally true.

The differential form of the magnetic circuit law

In Chapter 1 we saw that it is possible to express Gauss' theorem in either an integral form (Equation (1.5)) or a differential form (Equation (1.9)). The differential form turned out to be the more useful of the two in that it enabled a wider range of problems to be solved. The same is true of the magnetic circuit law, as we shall see. To derive the differential form of Equation (8.5) we apply it to the small rectangular path *ABCD* shown in Fig. 8.2. The surface *S* is taken to be in the *y*–*z* plane. The line integral of $\boldsymbol{H}$ is taken around *ABCD* in a direction which is in the right-handed corkscrew sense with respect to the positive *x*-direction. In other words, the elementary loop is encircled in the direction *ABCD*. The value of the line integral of $\boldsymbol{H}$ around the loop is computed by considering each side in turn. The effect of the dot product of $\boldsymbol{H}$ with $\mathbf{d}\boldsymbol{l}$ is to pick out the component of $\boldsymbol{H}$ which is parallel to the path. Thus, for *AB* we get

$$\int_A^B \boldsymbol{H}\cdot\mathbf{d}\boldsymbol{l} = H_y\delta y \tag{8.6}$$

Along *CD* the *y*-component of $\boldsymbol{H}$ can be written $(H_y + (\partial H_y/\partial z)\,\delta z)$, so that

$$\int_C^D \boldsymbol{H}\cdot\mathbf{d}\boldsymbol{l} = (H_y + (\partial H_y/\partial z)\,\delta z)\,(-\delta y) \tag{8.7}$$

The minus sign on the right-hand side of this equation is there because the path *CD* is traversed in a direction opposite to the positive *y*-direction. Combining Equations (8.6) and (8.7) gives

$$\int_A^B \boldsymbol{H}\cdot\mathbf{d}\boldsymbol{l} + \int_C^D \boldsymbol{H}\cdot\mathbf{d}\boldsymbol{l} = -\frac{\partial H_y}{\partial z}\,\delta y\,\delta z \tag{8.8}$$

Applying the same argument to the other two sides of the loop gives

$$\int_B^C \boldsymbol{H}\cdot\mathbf{d}\boldsymbol{l} + \int_D^A \boldsymbol{H}\cdot\mathbf{d}\boldsymbol{l} = \frac{\partial H_z}{\partial y}\,\delta y\,\delta z \tag{8.9}$$

The complete line integral around the loop *ABCD* is found by adding Equations (8.8) and (8.9) to give

$$\oint \boldsymbol{H}\cdot\mathbf{d}\boldsymbol{l} = \left(\frac{\partial H_z}{\partial y} - \frac{\partial H_y}{\partial z}\right)\delta y\,\delta z \tag{8.10}$$

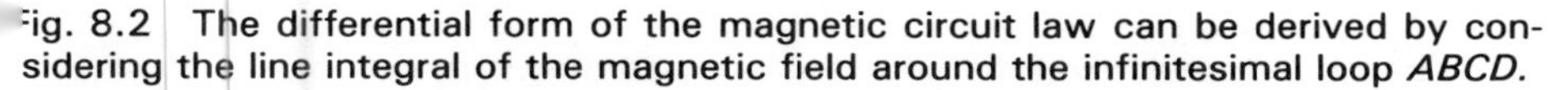

Fig. 8.2 The differential form of the magnetic circuit law can be derived by considering the line integral of the magnetic field around the infinitesimal loop *ABCD*.

Since the loop is small, it is reasonable to assume that J_x and D_x are constant over it, so that the integral on the right-hand side of Equation (8.5) becomes

$$\iint \left(\boldsymbol{J} + \frac{\partial \boldsymbol{D}}{\partial t} \right) \cdot \mathbf{d}\boldsymbol{A} = \left(J_x + \frac{\partial D_x}{\partial t} \right) \delta y \, \delta z \qquad (8.11)$$

This time the dot product picks out the components of $\boldsymbol{J}$ and $\boldsymbol{D}$ which are normal to the plane of the loop because the vector area $\mathbf{d}\boldsymbol{A}$ is also normal to the plane of the loop. Equating Equations (8.10) and (8.11) yields the differential form of Equation (8.5) for the x-direction:

$$\frac{\partial H_z}{\partial y} - \frac{\partial H_y}{\partial z} = J_x + \frac{\partial D_x}{\partial t} \qquad (8.12)$$

Similar equations can be obtained for the other two coordinate directions:

$$\frac{\partial H_x}{\partial z} - \frac{\partial H_z}{\partial x} = J_y + \frac{\partial D_y}{\partial t} \qquad (8.13)$$

and $$\frac{\partial H_y}{\partial x} - \frac{\partial H_x}{\partial y} = J_z + \frac{\partial D_z}{\partial t} \qquad (8.14)$$

Equations (8.12)–(8.14) together form the differential version of Equation (8.5) expressed in terms of the components of the vectors. As usual, it is possible to write these equations in a more compact form by using vector notation. To do this we multiply each by the appropriate unit vector and add them together to give

The vectors $\hat{\boldsymbol{x}}$, $\hat{\boldsymbol{y}}$ and $\hat{\boldsymbol{z}}$ are unit vectors in the x-, y- and z-directions.

$$\left(\frac{\partial H_z}{\partial y} - \frac{\partial H_y}{\partial z} \right) \hat{\boldsymbol{x}} + \left(\frac{\partial H_x}{\partial z} - \frac{\partial H_z}{\partial x} \right) \hat{\boldsymbol{y}} + \left(\frac{\partial H_y}{\partial x} - \frac{\partial H_x}{\partial y} \right) \hat{\boldsymbol{z}}$$
$$= \boldsymbol{J} + \frac{\partial \boldsymbol{D}}{\partial t} \qquad (8.15)$$

The left-hand side of this equation can be written as a determinant:

$$\begin{vmatrix} \hat{\boldsymbol{x}} & \hat{\boldsymbol{y}} & \hat{\boldsymbol{z}} \\ \dfrac{\partial}{\partial x} & \dfrac{\partial}{\partial y} & \dfrac{\partial}{\partial z} \\ H_x & H_y & H_z \end{vmatrix} \qquad (8.16)$$

which can be recognized as the cross product of the vector operator ∇ and the vector $\boldsymbol{H}$. Thus Equation (8.15) can be written succinctly in the form

The operation $\nabla \wedge \boldsymbol{H}$ is also known as curl $\boldsymbol{H}$.

$$\nabla \wedge \boldsymbol{H} = \boldsymbol{J} + \frac{\partial \boldsymbol{D}}{\partial t} \qquad (8.17)$$

Like the other differential forms derived earlier, Equation (8.17) is actually valid for coordinate systems other than the rectangular Cartesian system used to derive it.

The differential form of Faraday's law

The general integral form of Faraday's law was shown to be

$$\oint \boldsymbol{E} \cdot \mathbf{d}\boldsymbol{l} = -\iint \frac{\partial \boldsymbol{B}}{\partial t} \cdot \mathbf{d}\boldsymbol{A} \tag{8.18}$$

Equation (6.18)) for the case when the circuit is fixed in space and the magnetic lux density is changing with time. So far we have assumed that the line integral)f E is to be taken around a loop of wire. But if the loop is open-circuited and he potential difference between its ends is measured with a high-impedance volt-neter, then the loop can be thought of as a device for measuring the value of the ine integral of an electric field which exists in space whether the wire is present or not. Thus Equation (8.18) can be regarded as a generalization of Faraday's law, mplying that, if there is a changing magnetic field in any region of space, then here is also an electric field there. This is, strictly speaking, a plausible guess ather than a deduction, but it is one whose validity has been demonstrated by he correctness of the results derived from it. Making a comparison between Equations (8.5) and (8.18) allows us to deduce straight away that the differential orm of (8.18) must be

It is a good thing to remind yourself from time to time that all the theories of physical science are deductions from experimental observations and that, from the engineering point of view, no theory is much use until it has been validated by comparison with experiment.

$$\nabla \wedge \boldsymbol{E} = -\frac{\partial \boldsymbol{B}}{\partial t} \tag{8.19}$$

laxwell's equations

Ve have now introduced all the principles of electromagnetism and it is convenient o gather the main results together. Taking the integral forms first they are:

Gauss' theorem in electrostatics (Equation (2.5))

$$\oiint \boldsymbol{D} \cdot \mathbf{d}\boldsymbol{A} = \iiint \rho \, \mathrm{d}v \tag{8.20}$$

Gauss' theorem in magnetostatics (Equation (4.16))

$$\oiint \boldsymbol{B} \cdot \mathbf{d}\boldsymbol{A} = 0 \tag{8.21}$$

'he magnetic circuit law (Equation (8.5))

$$\oint \boldsymbol{H} \cdot \mathbf{d}\boldsymbol{l} = \iint \left(\boldsymbol{J} + \frac{\partial \boldsymbol{D}}{\partial t} \right) \cdot \mathbf{d}\boldsymbol{A} \tag{8.22}$$

'araday's law of induction (Equation (6.18))

$$\oint \boldsymbol{E} \cdot \mathbf{d}\boldsymbol{l} = -\iint \frac{\partial \boldsymbol{B}}{\partial t} \cdot \mathbf{d}\boldsymbol{A} \tag{8.23}$$

'he corresponding differential forms are

2.6)	$\nabla \cdot \boldsymbol{D} = \rho$	(8.24)
4.16)	$\nabla \cdot \boldsymbol{B} = 0$	(8.25)
8.17)	$\nabla \wedge \boldsymbol{H} = \boldsymbol{J} + \partial \boldsymbol{D}/\partial t$	(8.26)
8.19)	$\nabla \wedge \boldsymbol{E} = -\partial \boldsymbol{B}/\partial t$	(8.27)

'hese four equations form a summary of the whole of fundamental electro-agnetic theory. They are known, collectively, as **Maxwell's equations** and are the starting point for the discussion of all the more advanced topics in electromagnetism. In order to make use of them we also need a number of other equations which have been introduced in previous chapters. First there is the continuity equation

$$(3.14) \qquad \nabla \cdot \boldsymbol{J} = -\frac{\partial \rho}{\partial t} \tag{8.28}$$

Then there are the so-called **constitutive relations** which introduce the properties of materials, albeit in an idealized form. These are:

$$(2.4) \qquad \boldsymbol{D} = \epsilon \boldsymbol{E} \tag{8.29}$$

$$(3.3) \qquad \boldsymbol{J} = \sigma \boldsymbol{E} \tag{8.30}$$

$$(5.5) \qquad \boldsymbol{B} = \mu \boldsymbol{H} \tag{8.31}$$

The application of Equations (8.24)–(8.31) is most easily accomplished by making use of the methods of vector calculus. These techniques lie beyond the scope of this book, but a simple example of the use of Maxwell's equations which can be dealt with using more elementary mathematical methods is the subject of the next section.

Plane electromagnetic waves in free space

If this sounds far fetched it may help to imagine what would happen if a wave were propagated along a coaxial line whose outer conductor increased slowly in diameter with distance along the line.

In the previous chapter we saw that electromagnetic waves can propagate along a coaxial line and that the directions of the electric field vector, the magnetic field vector, and the direction of propagation are all mutually perpendicular. Taking our clue from this, we can see whether Maxwell's equations indicate that similar waves can propagate in free space far from any material boundaries. Let us assume that the wave propagates in the z-direction, and that the electric field has a component only in the x-direction. We assume, furthermore, that the intensity of the electric field varies only in the z-direction. Then from Equation (8.27), making use of the definition of the curl of a vector given in Equation (8.16), we have

$$\hat{y}\frac{\partial E_x}{\partial z} = -\frac{\partial \boldsymbol{B}}{\partial t} \tag{8.32}$$

since all the other components of $\nabla \wedge \boldsymbol{E}$ are zero. This equation shows that the magnetic field vector must lie in the y-direction, at right angles to both the electric field and the direction of propagation. Thus Equation (8.32) can be written as a scalar equation

$$\frac{\partial E_x}{\partial z} = -\frac{\partial B_y}{\partial t} \tag{8.33}$$

In free space the conduction current $\boldsymbol{J} = \boldsymbol{0}$, so Equation (8.26) becomes

$$-\frac{\partial H_y}{\partial z} = \frac{\partial D_x}{\partial t} \tag{8.34}$$

The magnetic field terms can be eliminated between these two equations by differentiating the first with respect to z and the second with respect to t, and b making use of Equations (8.29) and (8.31) with $\epsilon = \epsilon_0$ and $\mu = \mu_0$. The result is that

$$\frac{\partial^2 E_x}{\partial z^2} = -\mu_0 \frac{\partial^2 H_y}{\partial z\, \partial t} = \epsilon_0 \mu_0 \frac{\partial^2 E_x}{\partial t^2} \tag{8.35}$$

That is

$$\frac{\partial^2 E_x}{\partial z^2} = \frac{1}{v_{\mathrm{p}}^2} \frac{\partial^2 E_x}{\partial t^2} \tag{8.36}$$

where

$$v_{\mathrm{p}} = 1/\sqrt{(\epsilon_0 \mu_0)} \tag{8.37}$$

is the velocity of light. Equation (8.36) is the wave equation which was first encountered in Equation (7.4). If the electric field terms are eliminated from Equations (8.33) and (8.34) in a similar manner, the result is

$$\frac{\partial^2 H_y}{\partial z^2} = \frac{1}{v_{\mathrm{p}}^2} \frac{\partial^2 H_y}{\partial t^2} \tag{8.38}$$

Maxwell put forward the idea that light might be an electromagnetic phenomenon in 1873 but the experimental proof did not come until Hertz's experiments in 1888.

We observe that, according to Equation (7.48), the waves described by Equations (8.36) and (8.38) have phase velocities equal to the speed of light. The two wave equations have the following general solutions for waves travelling in the positive z-direction:

$$E_x = E_0 \exp[\mathrm{j}(\omega t - kz)] \tag{8.39}$$

and

$$H_y = H_0 \exp[\mathrm{j}(\omega t - kz)] \tag{8.40}$$

where E_0 and H_0 are complex amplitudes which incorporate the relative phases of the electric and magnetic fields. The relationship between the two fields can be found by substituting Equations (8.39) and (8.40) into either Equation (8.33) or Equation (8.44). Making use of Equation (8.33), we get

$$-\mathrm{j}kE_0 \exp[\mathrm{j}(\omega t - kz)] = -\mathrm{j}\omega\mu_0 H_0 \exp[\mathrm{j}(\omega t - kz)] \tag{8.41}$$

Remember that it is only the real parts of Equations (8.39) and (8.40) which have physical significance. The variations of E_x and H_y are sinusoidal in both time and space, as shown in Fig. 8.3.

From this it can be seen that E_x and H_y are in phase with one another. Also

$$E_0/H_0 = \sqrt{(\mu_0/\epsilon_0)} = 377\ \Omega \tag{8.42}$$

We recognise this as the wave impedance of a wave in free space (see Equation (7.51)). It is sometimes referred to as the **intrinsic impedance of free space.** Figure 8.3 illustrates the complete solution which has been obtained. It is necessary to be a little careful in interpreting this diagram. It shows the amplitudes of the field vectors rather than being a map of the field. The whole field pattern is moving in the z-direction with the velocity of light. At any instant the electric and magnetic field strengths are uniform at all points in a plane perpendicular to the z-axis.

Power flow in an electromagnetic wave

In Chapter 7 it was shown that the power flow in a coaxial line could be calculated correctly by assuming that the stored energy in the electromagnetic field travelled

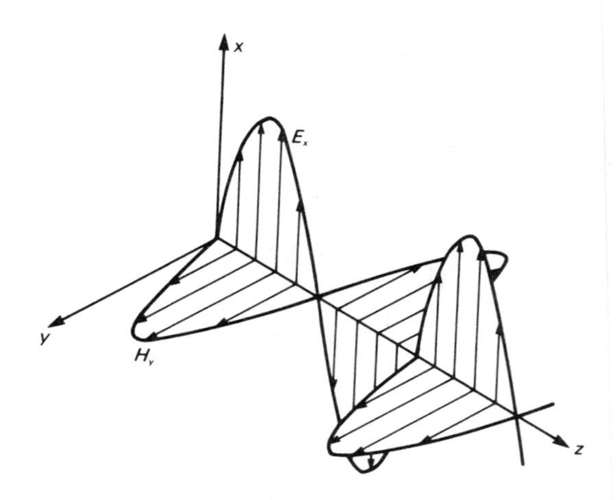

Fig. 8.3 The electric and magnetic fields in a plane electromagnetic wave propagating in the z-direction.

with the phase velocity in the direction of the wave. For waves in free space the energy density is, from Equations (2.10) and (6.37),

$$\begin{aligned} w &= \tfrac{1}{2}\boldsymbol{D}\cdot\boldsymbol{E} + \tfrac{1}{2}\boldsymbol{B}\cdot\boldsymbol{H} \\ &= \tfrac{1}{2}\epsilon_0 E_x^2 + \tfrac{1}{2}\mu_0 H_y^2 \\ &= \epsilon_0 E_x^2 \end{aligned} \tag{8.43}$$

The instantaneous power density in the wave is

$$\begin{aligned} P &= w\, v_{\mathrm{p}} \\ &= \epsilon_0 E_x^2/\sqrt{(\epsilon_0\mu_0)} \\ &= E_x H_y \end{aligned} \tag{8.44}$$

Now P is in the z-direction, so it is possible to write Equation (8.44) in vector notation as

$$\boldsymbol{P} = \boldsymbol{E} \wedge \boldsymbol{H} \tag{8.45}$$

The vector $\boldsymbol{P}$ is known as the **Poynting vector**. Poynting's theorem states:

> The integral of $(\boldsymbol{E} \wedge \boldsymbol{H})$ over a closed surface is equal to the instantaneous flow of electromagnetic power out of the volume enclosed by that surface.

The proof of this theorem is beyond the scope of this book. Note that, strictly, it is the integral of $\boldsymbol{P}$ over a closed surface which has meaning rather than $\boldsymbol{P}$ itself. We have already seen that the integral of the Poynting vector gives the correct answer for the power flow in a lossless coaxial line. We will now look at another example of its use.

Worked Example 8.1 A long straight cylindrical wire carries a current I. Given that the wire has radius a and resistance R per unit length, calculate the power dissipated per unit length both directly and by integration of the Poynting vector.

Solution (a) *By direct methods*. The power dissipated per unit length is I^2R.

(b) *By Poynting's theorem*. Just outside the surface of the wire, the magnetic circuit law gives $2\pi a H_\theta = I$, so that $H_\theta = I/2\pi a$.

The voltage drop per unit length is $V = IR = E_z$ (the electric field just outside the wire). Integrating the Poynting vector over a cylinder just outside the wire which has radius a and unit length gives the total power flow into the wire as

$$E_z H_\theta (2\pi a) = (I/2\pi a)(IR)(2\pi a) = I^2 R$$

as before.

The figure in the margin shows the direction of E_z and H_θ. It is clear that the Poynting vector is directed into the surface of the wire. Thus, using the field picture, the ohmic power loss in the wire results in a steady flow of electromagnetic energy into the wire from the surrounding field.

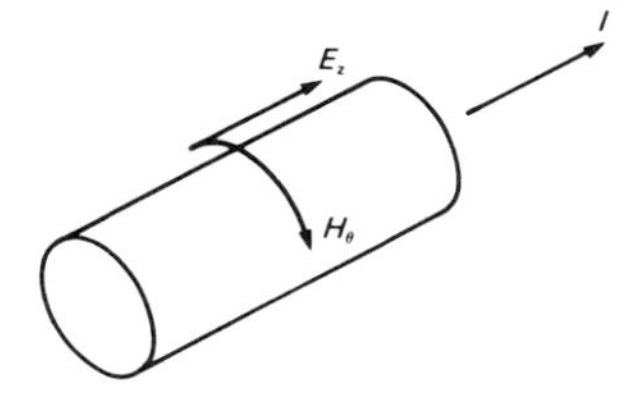

Summary

We have considered how Maxwell removed an inconsistency in the magnetic circuit law by introducing the idea of the displacement current. The modified law was expressed in differential form by introducing the curl of the vector ***H***. Faraday's law was also put into differential form by using the curl of ***E***. The mathematical statements of the laws of electromagnetism were collected together in both their integral and differential forms. This set of equations is known as Maxwell's equations. It was demonstrated that, for the special case of plane waves in free space, Maxwell's equations lead naturally to the prediction of the existence of electromagnetic waves which travel with the speed of light. Finally, it was shown that the power flow in the electromagnetic field can be represented by the Poynting vector $(\boldsymbol{E} \wedge \boldsymbol{H})$.

9 Screening circuits against radio-frequency interference

Objectives

- ☐ To introduce the idea that small electric and magnetic dipoles can radiate and pick up electromagnetic waves.
- ☐ To show that electromagnetic waves can only penetrate a short distance into conducting materials.
- ☐ To discuss the screening effectiveness of metal enclosures at radio frequencies.
- ☐ To show that a small hole in a conducting sheet behaves as an electric or magnetic dipole.
- ☐ To estimate the reduction in the screening effectiveness of an enclosure caused by leakage of radiation through a small hole in it.
- ☐ To introduce the idea of electromagnetic resonances in conducting enclosures.
- ☐ To estimate the reduction in the screening effectiveness of an enclosure caused by a resonance.

A major concern of a designer of electronic equipment must be to ensure that its operation is not affected by external electromagnetic fields. A related concern is to ensure that it does not affect other equipment nearby by radiating electromagnetic energy. The problem of ensuring that electronic equipment is neither susceptible to nor causes electromagnetic interference is known as **electromagnetic compatibility** (emc). This topic is of growing importance and is the subject of national and international standards.

We have already noted that electromagnetic interference can be caused at low frequencies by parasitic capacitive or inductive coupling. These effects can be reduced or eliminated by careful circuit design and by the use of electric and magnetic screens. At higher frequencies emission of and susceptibility to radiated electromagnetic energy become important. Sources of radiated interference include intentional radiators (radio transmitters, portable telephones), unintentional radiators (radio receivers, digital electronic circuits), incidental radiators (electric motors, car ignition systems, fluorescent lights) and natural sources (atmospheric interference, lightning). The frequency spectra of these sources lie in the range from 10 kHz to 1 GHz or higher. The problems of screening against radiofrequency interference are greater than at lower frequencies. The purpose of this chapter is to emphasize the physical principles involved and to introduce ways of making estimates of screening effectiveness. The treatment is simplified and intended to provide an introduction to the detailed discussions to be found in the specialized texts listed in the Bibliography.

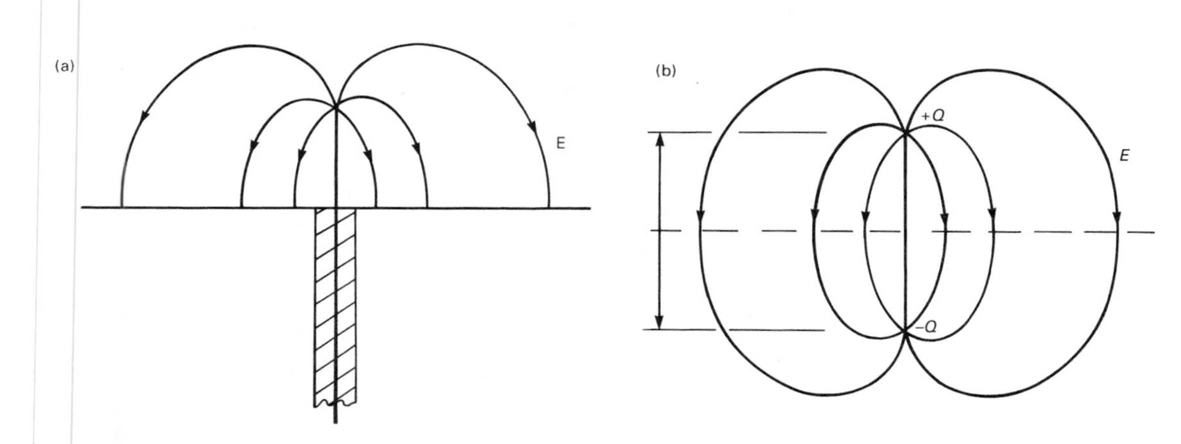

Fig. 9.1 The electric fields around (a) the end of a coaxial line, and (b) an electric dipole.

The small electric dipole

We have seen that the signal propagating down a coaxial line can be regarded as an electromagnetic wave. If the line is air-spaced then the phase velocity and wave impedance of the wave are the same as those of a wave propagating in free space. It is reasonable, therefore, to suppose that, if a line is terminated by an open end, some of the energy flowing down the line may be radiated into space. Figure 9.1(a) shows a coaxial line which is terminated by connecting the outer conductor to a conducting plane while the inner conductor protrudes a short distance into the space beyond. The termination is likely to approximate to an open circuit, so there will be a voltage standing-wave maximum at the end of the line. The electric field pattern must be somewhat as shown. By using the method of images we find that the field pattern is just one-half of that of the isolated **electric dipole** shown in Fig. 9.1(b).

This field pattern is most conveniently described in the system of **spherical polar coordinates** shown in Fig. 9.2. In this figure r is the length of OP, θ is the angle between OP and the z-axis, and ϕ is the angle between the projection of OP on the x–y plane and the x-axis. Although the coordinate directions are not fixed in space they are always at right angles to each other as shown in the inset to Fig. 9.2. If the direction of the dipole is chosen to coincide with the z-axis then the fields must be circularly symmetric about that axis. The analysis of this problem is beyond the scope of this book. Here we shall merely quote the results and examine their implications for electromagnetic coupling.

A fuller discussion of the field of an oscillating dipole is to be found in R.G. Carter.*

It is convenient to distinguish two different regions. In the near field of the dipole $r \ll \lambda/2\pi$. Now the free-space wavelength is 300 mm at 1 GHz and 300 m at 1 MHz, so this condition is satisfied for coupling by radiation between two parts of the same piece of electronic equipment.

The components of the electric and magnetic fields in this region are

$$E_r = \frac{p\cos\theta}{2\pi\epsilon_0 r^3}\exp(j\omega t) \tag{9.1}$$

$$E_\theta = \frac{p\sin\theta}{4\pi\epsilon_0 r^3}\exp(j\omega t) \tag{9.2}$$

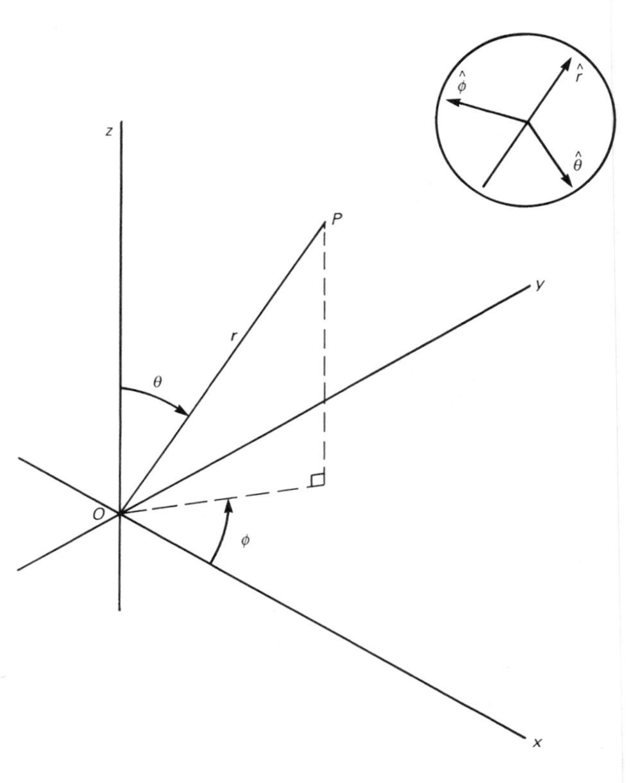

Fig. 9.2 Spherical polar coordinates.

$$H_\phi = \mathrm{j}\frac{\omega p \sin\theta}{4\pi r^2}\exp(\mathrm{j}\omega t) \tag{9.3}$$

where $p = Ql$ is the **electric dipole moment**. We note that the directions of these fields are as we would expect from the symmetry of the problem. It can be shown that the electric field is just the field of an electrostatic dipole and that the magnetic field arises directly from the Biot–Savart law (Equation (4.6)). The phase difference of 90° between the electric and magnetic fields means that there is no net flow of energy. The wave impedance (E_θ/H_ϕ) in this region is greater than the wave impedance in free space.

In the equivalent circuit of the dipole these fields are represented by a capacitance. This capacitance can be estimated by approximating the dipole by two spheres of radius a, whose centres are a distance d apart, joined by a thin wire. The charges will then reside almost entirely on the spheres and the capacitance can be calculated using the method of images in the same way as in Worked Example 2.4. If $a \ll d$ the capacitance is given by

$$C = 2\pi\epsilon_0\left[\frac{1}{a} - \frac{1}{d}\right]^{-1} \tag{9.4}$$

The full solution for the dipole fields contains additional terms which vary as $1/r^2$. These are significant at distances from the dipole of the order of $\lambda/2\pi$.

Strictly speaking, there should be an inductive component as well, but if $1 \ll \lambda$ then this is negligible compared with the capacitance.

In the far-field region $r > \lambda$. This is the situation which exists when a piece of electronic equipment is illuminated by radiation from a remote source. The field components are

$$E_\theta = -\frac{k_0^2 p \sin\theta}{4\pi\epsilon_0 r} \exp[\mathrm{j}(\omega t - k_0 r)] \tag{9.5}$$

$$H_\phi = -\frac{\omega k_0 p \sin\theta}{4\pi\epsilon_0 r} \exp[\mathrm{j}(\omega t - k_0 r)] \tag{9.6}$$

These are in phase with each other and so represent a net flow of electromagnetic energy. From the inset of Fig. 9.2 and Equation (8.45) we can see that the energy flow is radially outwards. The power density is zero along the z-axis and maximum in the x–y plane. The ratio of the amplitudes of E_θ to H_ϕ is equal to the free-space wave impedance given by Equation (8.42).

The Poynting vector can be integrated over the surface of a sphere surrounding the dipole to give the total power radiated:

$$P = \frac{\omega^2 k_0^2 p^2 Z_w}{12\pi} \tag{9.7}$$

This expression is independent of the radius of the sphere because the power flow is the same for any closed surface surrounding the dipole. Now, since the current in the dipole is

$$I = \frac{\mathrm{d}Q}{\mathrm{d}t} = \mathrm{j}\omega Q \tag{9.8}$$

we can write Equation (9.7) in the form

$$P = 40\pi^2 I^2 (l/\lambda)^2 \tag{9.9}$$

The input resistance of the dipole is therefore

$$R_r = \frac{2P}{I^2} = 80\pi^2 (l/\lambda)^2 \tag{9.10}$$

In all situations likely to be important for electromagnetic compatibility $l \ll \lambda$ so this resistance is a few ohms at most. Figure. 9.3(a) shows the equivalent circuit of the dipole. As the dipole approximates to an open circuit the reactance X_r must be large.

An electric dipole can also act as a receiving antenna and extract power from an electromagnetic wave incident upon it. Figure 9.3(b) shows this situation. The voltage induced in the dipole is $E_i l$ with an effective source impedance of $(R_r + \mathrm{j}X_r)$. Thus the equivalent circuit of the diode connected to a load is as shown in Fig. 9.3(c). This circuit can be used to estimate the voltage appearing at the load as a result of electromagnetic coupling if the dipole reactance can be estimated or measured.

Worked Example 9.1

A piece of wire 1 mm in diameter and 10 mm long is left protruding from the surface of a printed circuit board. The end of the wire is connected to the input of an amplifier whose input impedance is 1 MΩ. Assuming that the electromagnetic field strength associated with man-made electrical noise is 10 μV m^{-1}

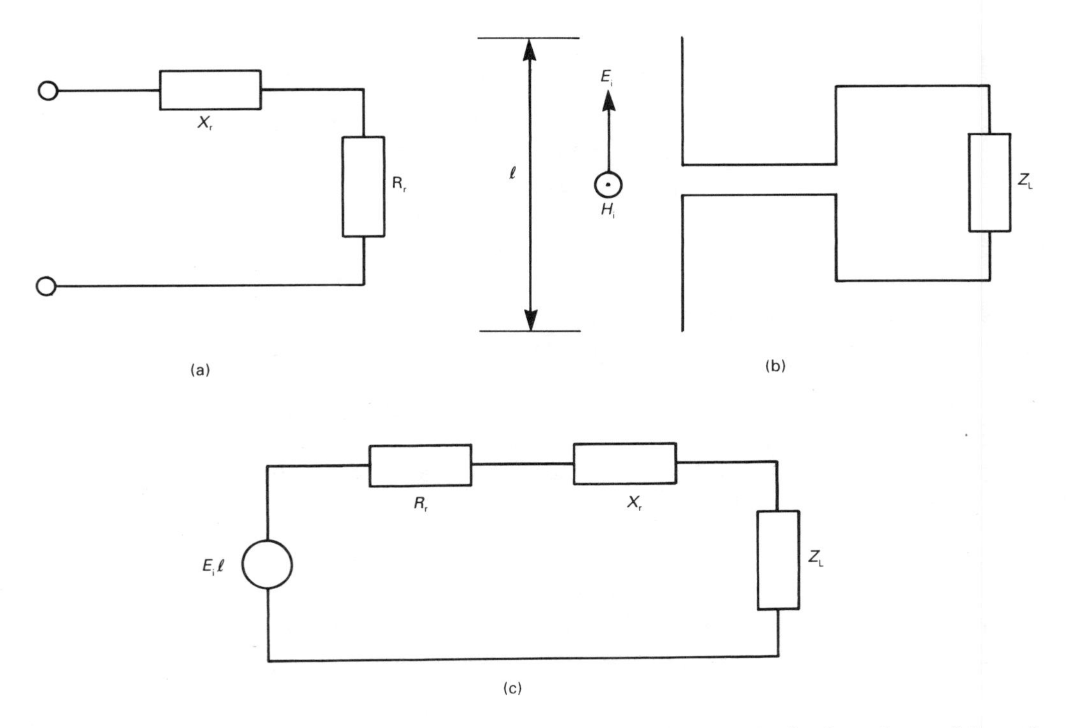

Fig. 9.3 (a) The equivalent circuit of an electric dipole radiator; (b) an electric dipole acting as a receiving antenna and (c) its equivalent circuit.

Some useful information about sources of radiated interference is given by Keiser*.

at 10 MHz (a figure typical of the field strength produced by a radio transmitter in its primary service area), estimate the amplitude of the spurious signal at the input of the amplifier.

Solution The voltage induced in the piece of wire is $10 \times 10^{-2} = 0.1$ mV. The wavelength at 10 MHz is 30 m so, from Equation (9.10), the dipole source resistance is 0.09 mΩ. If we assume that $a = 1$ mm and $d = 20$ mm in Equation (9.4) then the capacitance of a complete dipole is 0.06 pF. The capacitance of the piece of wire is twice this so the dipole reactance is of the order of 133 kΩ. Thus the dipole source impedance is appreciably less than the amplifier input impedance and the amplitude of the spurious signal will be of the order of 0.1 mV.

Note that we have assumed that the wire is parallel to the direction of the electric field in the incident wave because this is the worst case. If the wire were perpendicular to the electric field then the induced voltage would be zero.

The small magnetic dipole

Details of the properties of the small magnetic dipole are to be found R.G. Carter.*

The other type of elementary radiating element is the magnetic dipole. This takes the form of a closed loop of wire carrying a radiofrequency current as shown in Fig. 9.4(a). Loops are a common feature of electronic circuits. They are formed, for example, when components are mounted on printed circuit boards, as shown in Fig. 9.4(b). Figure 9.4(a) shows that, sufficiently far from the loop, the pattern

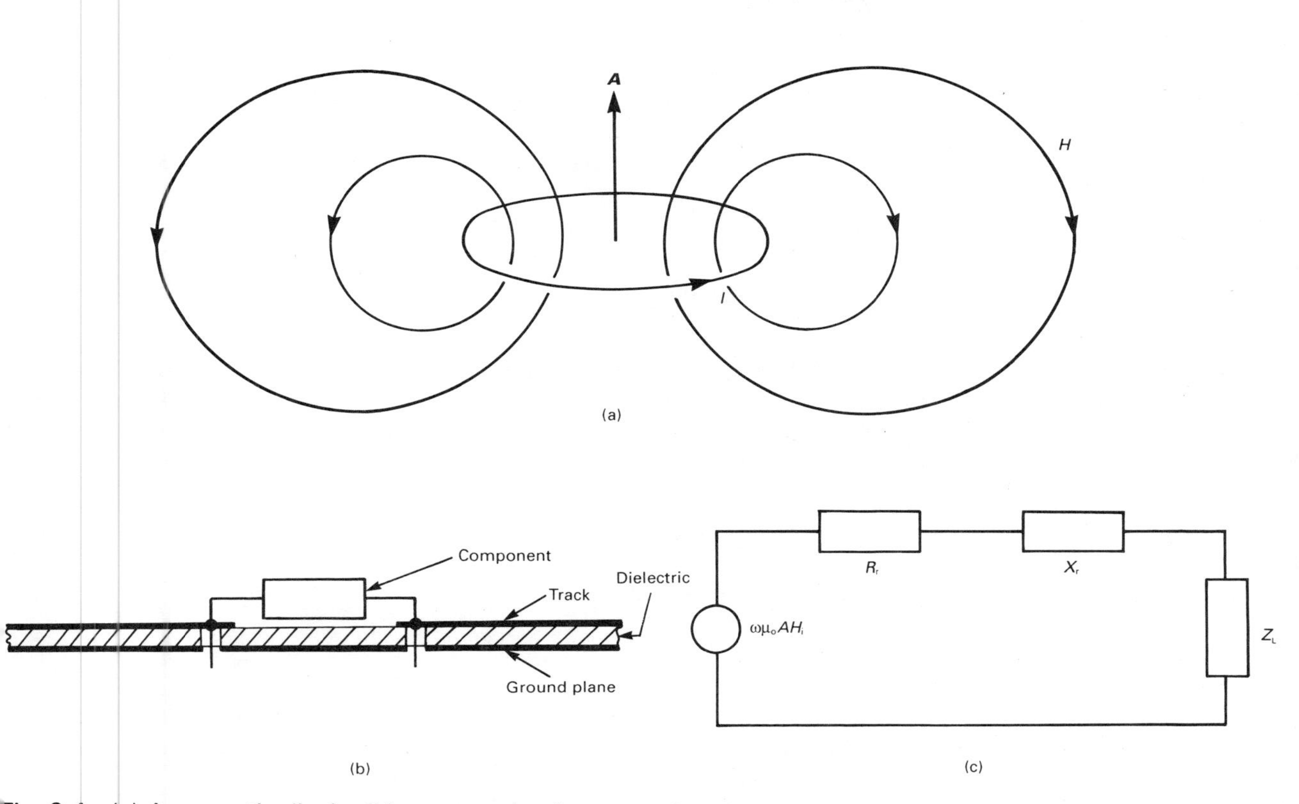

Fig. 9.4 (a) A magnetic dipole; (b) an example of a magnetic dipole antenna in an electronic circuit and (c) its equivalent circuit.

of the magnetic field is the same as that of the electric field of the electric dipole. It can be shown that the two dipoles are duals of each other. If the loop lies in the x–y plane and has area A then the near-field components are

$$H_r = \frac{j_m \cos\theta}{2\pi r^3} \exp(j\omega t) \tag{9.11}$$

$$H_\theta = \frac{j_m \sin\theta}{4\pi r^3} \exp(j\omega t) \tag{9.12}$$

$$E_\phi = -j\frac{\omega\mu_0 j_m}{4\pi r^2} \exp(j\omega t) \tag{9.13}$$

where $j_m = AI$ is the **magnetic dipole moment**. The wave impedance is less than the wave impedance of free space in this region. This implies that the coupling is caused chiefly by the magnetic field and takes the form of mutual inductance.

These fields store energy and are represented in the equivalent circuit of the dipole by an inductance. The exact calculation of the self-inductance of a loop of wire is difficult, but a useful approximate formula is given by Walker* for a circular loop of radius R made of wire whose radius is r.

$$L \simeq \mu_0 R\left[\ln\left(\frac{8R}{r}\right) - 2\right] \tag{9.14}$$

This equation is accurate enough for our purposes. It provides a reasonable estimate of the self-inductance of a non-circular loop if R is the radius of a circular loop having the same area.

The far field of the loop is given by

$$H_\theta = -\frac{j_m k_0^2 \sin\theta}{4\pi r} \exp[j(\omega t - k_0 r)] \tag{9.15}$$

$$E_\phi = \frac{j_m \mu_0 \omega k_0 \sin\theta}{4\pi r} \exp[j(\omega t - k_0 r)] \tag{9.16}$$

Note that the roles of the electric and magnetic fields have been exchanged from those in the case of the electric dipole. The Poynting vector is directed radially outwards and the wave impedance is that of free space. For a circular loop of radius a the power radiated is

$$P = \frac{1}{12\pi}(Z_w k_0^4 j_m^2) \tag{9.17}$$

and the radiation resistance is

$$R_r = \frac{8\pi Z_w}{3}\left(\frac{\pi a}{\lambda}\right)^4 \tag{9.18}$$

Such a loop can also pick up radiated signals. If the magnetic field in the incident wave is H_i (assumed to be normal to the plane of the loop) then the amplitude of the induced e.m.f. is

$$\mathcal{E} = \omega \mu_0 A H_i \tag{9.19}$$

The equivalent circuit of the loop is as shown in Fig. 9.4(c).

Worked Example 9.2 Estimate the e.m.f. induced in a loop of wire 1 mm in thickness, whose area is 0.01 m^2, when it is placed in an electromagnetic field of 10 μV m^{-1} at a frequency of 10 MHz.

Solution To find the e.m.f. induced in the loop we first note that the amplitude of the magnetic field in the incident wave is $10^{-5}/377 = 0.027 \times 10^{-6}$ A m^{-1}. The flux linking the loop is greatest when the plane of the loop is normal to the direction of the magnetic field. It is then 0.34×10^{-15} Wb. We assume that the loop can be approximated by a circular loop having the same area. The radius of this loop is 56 mm. From Equation (9.18) the radiation resistance is 3.7 μΩ. The self-inductance of the loop is obtained from Equation (9.14) as 0.34 μH, so its reactance is 21 Ω. The induced e.m.f. is given by Equation (9.19) as 0.02 μV.

Electromagnetic waves in conducting materials

To investigate the propagation of electromagnetic waves in conducting materials we shall assume that the waves propagate in the z-direction and that the electric and magnetic fields only have components in the x- and y-directions, respectively. From Equation (8.27) we derive

$$\frac{\partial E_x}{\partial z} = -\frac{\partial B_y}{\partial t} \tag{9.20}$$

which is identical to Equation (8.33). From Equation (8.26) we obtain

$$-\frac{\partial H_y}{\partial z} = \frac{\partial D_x}{\partial t} + \sigma E_x \tag{9.21}$$

which differs from Equation (8.34) in the addition of the conduction-current term. If we now assume that all quantities vary as $\exp[\mathrm{j}(\omega t - kz)]$ the two equations above become

$$\mathrm{j}kE_x = \mathrm{j}\omega\mu_0 H_y \tag{9.22}$$

$$\mathrm{j}kH_y = (\mathrm{j}\omega\epsilon + \sigma)E_x \tag{9.23}$$

In a good conductor $\sigma \gg \mathrm{j}\omega\epsilon$ so Equation (9.21) may be approximated as

$$\mathrm{j}kH_y = \sigma E_x \tag{9.24}$$

When E_x and H_y are eliminated between Equations (9.20) and (9.21) the result is

$$k^2 = -\mathrm{j}\omega\sigma\mu \tag{9.25}$$

or

$$k = \sqrt{-\mathrm{j}}\,\sqrt{\omega\sigma\mu} \tag{9.26}$$

To evaluate $\sqrt{-\mathrm{j}}$ we note that

$$-\mathrm{j} = \exp(-\mathrm{j}\pi/2) \tag{9.27}$$

so that

$$\begin{aligned}\sqrt{-\mathrm{j}} &= \exp(-\mathrm{j}\pi/4) \\ &= \cos(-\pi/4) + \mathrm{j}\sin(-\pi/4) \\ &= (1 - \mathrm{j})/\sqrt{2}\end{aligned} \tag{9.28}$$

and

$$\begin{aligned}k &= \sqrt{\omega\sigma\mu/2}\,(1 - \mathrm{j}) \\ &= (1 - \mathrm{j})/\delta\end{aligned} \tag{9.29}$$

where

$$\delta = \sqrt{\frac{2}{\omega\sigma\mu}} \tag{9.30}$$

Thus waves propagate in a conducting material as $\exp[\mathrm{j}(\omega t - z/\delta)]\exp(-z/\delta)$, decaying exponentially to an amplitude 1/e in the **skin depth** δ. This is a very important result. It shows that waves are not completely excluded from conducting materials but penetrate a short distance into them.

Worked Example 9.3

What is the skin depth in aluminium at a frequencies of 50 Hz, 1 KHz and 1 MHz?

Solution The conductivity of aluminium is 3.5×10^7 S m^{-1}, so the skin depth at 50 Hz is $(0.5 \times 2\pi \times 50 \times 3.5 \times 10^7 \times \mu_0)^{-\frac{1}{2}} = 12.0$ mm. Similarly, at 1 kHz $\delta = 2.7$ mm, and at 1 MHz $\delta = 0.085$ mm.

If the skin depth is comparable with the radius of the wire the theory is more difficult. Further information is given by G.W. Carter*.

At high frequencies the **skin effect** means that an electric current does not flow through the body of a wire but only in a thin surface layer. If the skin depth is much less than the radius of the wire then we can open out the surface of the wire into a flat strip, as shown in Fig. 9.5. The current density is given by

$$J = J_0 \exp[-(1-\mathrm{j})z/\delta] \exp(\mathrm{j}\omega t) \tag{9.31}$$

where J_0 is the current density at the surface of the wire. The power dissipated in a unit length of the element dz is then

$$\begin{aligned} \mathrm{d}P &= \tfrac{1}{2}\rho |J|^2 w \,\mathrm{d}z \\ &= \tfrac{1}{2}\rho J_0^2 \exp(-2z/\delta) w \,\mathrm{d}z \end{aligned} \tag{9.32}$$

The total power dissipated per unit length of the strip is obtained by integrating Equation (9.32). The upper limit can be taken at infinity since the current falls to zero long before the centre of the wire is reached. Thus

$$\begin{aligned} P &= \tfrac{1}{2}\rho w J_0^2 \int_0^\infty \exp(-2z/\delta)\,\mathrm{d}z \\ &= \tfrac{1}{4}\rho w \delta J_0^2 \end{aligned} \tag{9.33}$$

Now the total current flowing in the strip is given by

$$\begin{aligned} I &= J_0 w \exp(\mathrm{j}\omega t) \int_0^\infty \exp[-(1-\mathrm{j})z/\delta]\,\mathrm{d}z \\ &= J_0 w \exp(\mathrm{j}\omega t) \left(\frac{-\delta}{1-\mathrm{j}}\right) \end{aligned} \tag{9.34}$$

It will be noticed that the phase of J varies with z and so the phase of I is not the same as that of J_0. The amplitude of the current is

$$|I| = J_0 w\delta/\sqrt{2} \tag{9.35}$$

so that equation (9.33) may be written

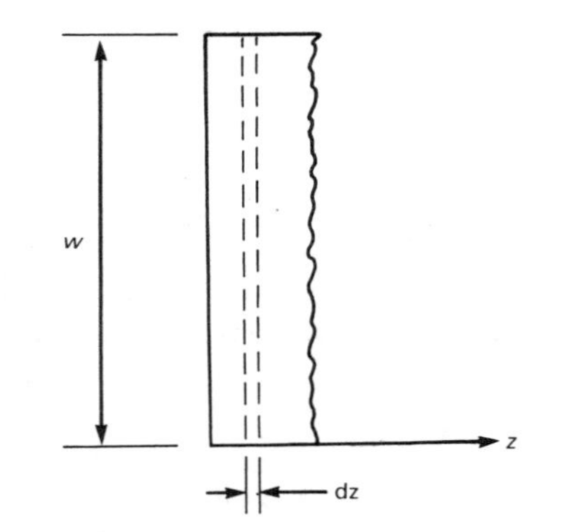

Fig. 9.5 The surface of a round wire opened out into a flat strip.

$$P = \tfrac{1}{2}\frac{\rho}{w\delta}|I|^2$$
$$= \tfrac{1}{2}R|I|^2 \qquad (9.36)$$

where $R = \rho/w\delta$ is the resistance per unit length of a layer of thickness δ at the surface of the wire. Thus at high frequencies the current flowing in a wire may be considered to flow uniformly in a surface layer whose thickness is equal to the skin depth. As a consequence, the resistance of a wire is not constant but increases with frequency.

Worked Example 9.4

What is the resistance of a copper wire 1 mm in diameter and 10 m long at low frequencies and at 1 MHz?

Solution The conductivity of copper is 5.7×10^7 S m^{-1}. At low frequencies the resistance is given by Equation (3.6). The result is 0.22 Ω. At 1 MHz the skin depth is 67 μm so the resistance is the same as that of a tube whose thickness is equal to the skin depth: 0.83 Ω.

Reflection of electromagnetic waves by a conducting surface

An important problem for electromagnetic screening is the reflection of electromagnetic waves from the surface of a conducting material. Figure 9.6 shows the boundary between air and a conductor with an electromagnetic wave (E_i, H_i) incident normally upon it. The wave transmitted into the conductor has field amplitudes (E_t, H_t). We must also assume that there may be a reflected wave with field amplitudes (E_r, H_r). Note that the directions assumed for the field vectors are consistent with wave propagation in the directions shown by the Poynting vectors S_1, S_t and S_r.

The electric and magnetic fields must be continuous at $z = 0$ (see pages 26 and 70), so that

$$E_i + E_r = E_t \qquad (9.37)$$

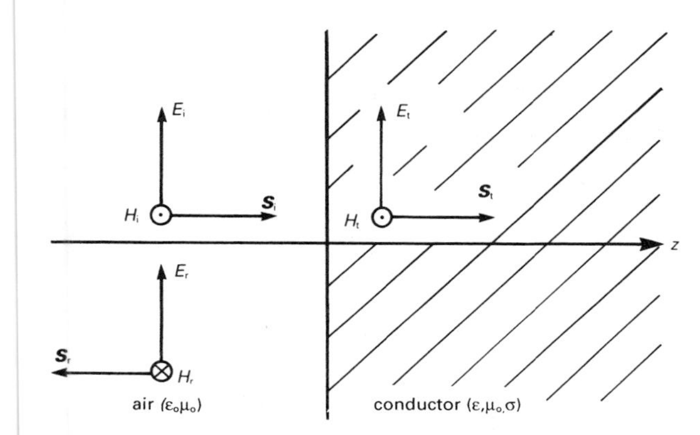

Fig. 9.6 Reflection of electromagnetic waves at the interface between air and a conducting material.

and

$$H_i - H_r = H_t \tag{9.38}$$

Now the electric and magnetic field amplitudes are not independent of each other. In air we have

$$E_i/H_i = E_r/H_r = \sqrt{\epsilon_0/\mu_0} = Z_1 \tag{9.39}$$

from Equation (8.42). In the conductor

$$E_t/H_t = jk/\sigma = (1 + j)/\sigma\delta = Z_2 \tag{9.40}$$

from Equations (9.24) and (9.29). Equation (9.38) may therefore be written

$$(E_i - E_r)/Z_1 = E_t/Z_2 \tag{9.41}$$

It is easiest to find the transmitted power by first calculating the reflected power. When E_t eliminated between equations (9.37) and (9.41) the result is

$$\frac{E_r}{E_i} = \frac{Z_2 - Z_1}{Z_2 + Z_1} \tag{9.42}$$

Now $Z_1 = 377\ \Omega$ in free space and Z_2 is typically very small. For example, for aluminium at 1 kHz, $Z_2 = (1 + j) \times 10.6 \times 10^{-6}\ \Omega$. Figure 9.7 shows the phasor diagram for adding Z_1 to Z_2. It is evident that to a good approximation

$$|Z_1 \pm Z_2| = |Z_1| \pm \frac{1}{\sqrt{2}}|Z_2| \tag{9.43}$$

so that

$$\begin{aligned}\left|\frac{E_r}{E_i}\right| &= \frac{|Z_1| - |Z_2|/\sqrt{2}}{|Z_1| + |Z_2|/\sqrt{2}} \\ &\simeq 1 - \sqrt{2}\,|Z_2/Z_1|\end{aligned} \tag{9.44}$$

The ratio of the transmitted power to the incident power is given by

$$\begin{aligned}\frac{P_t}{P_i} &= 1 - \left|\frac{E_r}{E_i}\right|^2 \\ &= 1 - [1 - 2\sqrt{2}\,|Z_2/Z_1|] \\ &= \frac{4}{Z_1\sigma\delta}\end{aligned} \tag{9.45}$$

Thus the attenuation of the signal by reflection at the surface of the conductor known as the **reflection loss**, may be written in decibels as

$$L = -10\log\left(\frac{4}{Z_1\sigma\delta}\right) \tag{9.46}$$

Similarly, the reflection loss for a signal passing from a conductor into air is given by

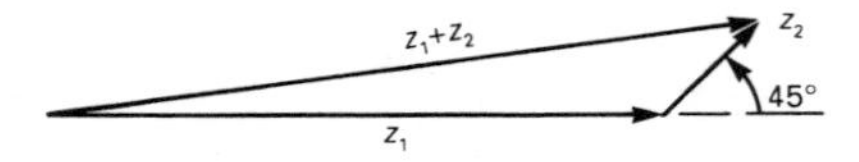

Fig. 9.7 The phasor addition of the wave impedances of air and of a conducting material.

$$L = -10\log\left(\frac{8}{Z_1\sigma\delta}\right) \tag{9.47}$$

The derivation of Equation (9.47) may be found in R.G. Carter.*

Now the signal is attenuated as $\exp(-z/\delta)$ as it passes through the conductor so the **transmission loss** is

$$\begin{aligned} L &= -20\log[\exp(-z/\delta)] \\ &= 8.69\,z/\delta \end{aligned} \tag{9.48}$$

Provided that the attenuation through the thickness of the material is sufficient for multiple reflections to be negligible, we can use Equations (9.46)–(9.48) to estimate the screening effectiveness of an enclosure.

Worked Example 9.5

Estimate the screening effectiveness of an enclosure made of aluminium sheet 0.5 mm thick at a frequency of 1 MHz.

Solution From Worked Example 9.3 the skin depth in aluminium is 0.085 mm at 1 MHz. Then from Equation (9.46) the reflection loss at the air–aluminium interface is 54 dB, from Equation (9.47) the reflection loss at the aluminium– air interface is 51 dB, and from Equation (9.48) the transmission loss through the aluminium is 51 dB. The last figure indicates that multiple reflections are not a problem in this case. The screening effectiveness is therefore the sum of the reflection and transmission losses: 156 dB.

In many cases the main contribution to the effectiveness of a screen is the mismatch between the wave impedances of the screen and of the incident wave. It is desirable that the mismatch should be as large as possible. Now in the near-field region the wave impedance of an electric dipole is greater than that of a magnetic dipole. Thus the screen is more effective against the field of the electric dipole than that of the magnetic dipole. In the limit of low frequencies a non-magnetic conducting screen has no ability to exclude magnetic fields.

Further information about the screening effectiveness of enclosures is to be found in Keiser* and in R.G. Carter.*

The effect of holes in screening enclosures

It would appear from the previous section that the screening effectiveness of a metal enclosure is so good that all problems of electromagnetic interference can be solved in this way. Unfortunately this is not true. The reason is that any practical piece of electronic equipment must exchange information with the outside world, and for that there must usually be one or more holes in the screen. Other problems can arise from the way in which the enclosure is made — good electrical contact is essential at all the joints. Electrical interference can be conveyed into or out of the enclosure either by conduction along wires or by radiation through apertures. Interference caused by conduction along wires is dealt with by the use of filters. In this section we consider the problem of radiation through holes in a screen.

Details of the techniques used to ensure that good shielding is obtained are given by Keiser*.

When an electromagnetic wave is incident on a conducting sheet the tangential component of the electric field and the normal component of the magnetic field

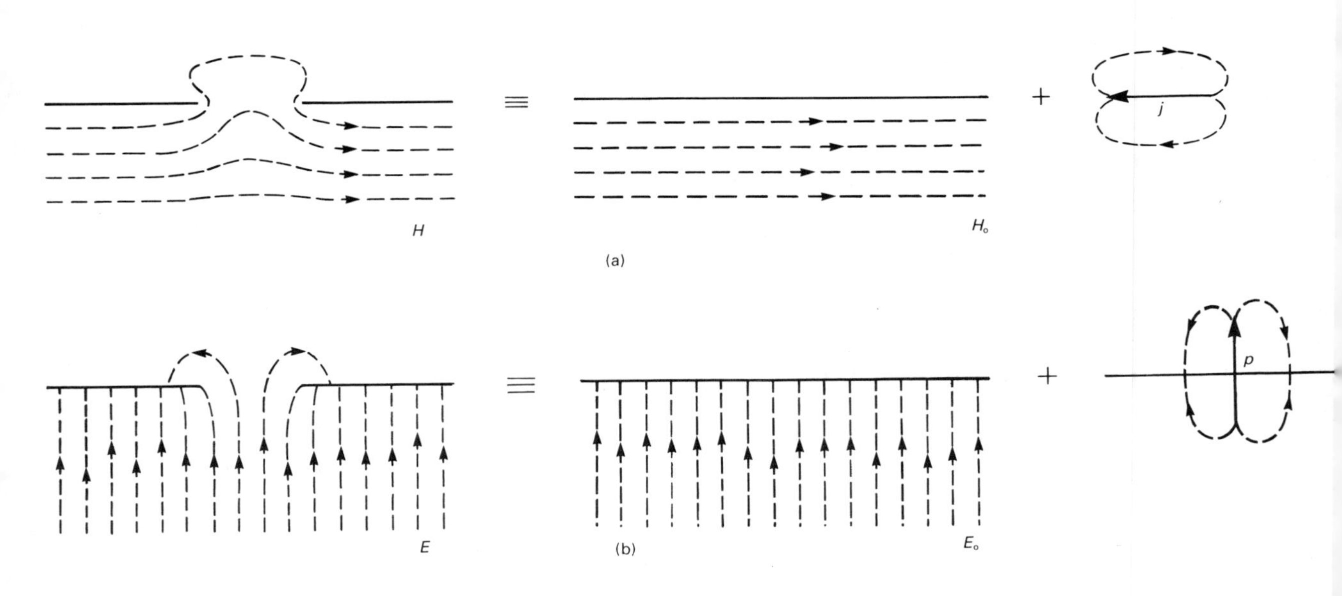

Fig. 9.8 The penetration of (a) a magnetic field and (b) an electric field through a hole in a conducting sheet represented by equivalent dipoles.

must always be zero. It follows that any general case can be considered as the superposition of the two special cases: tangential magnetic field and normal electric field. If there is a hole in the sheet then some of the field passes through it as shown in Fig. 9.8. Provided that the hole is small compared with the wavelength of the radiation, it is possible to use the static field solutions. The complete field patterns can be represented, as shown in Fig. 9.8, by the superposition of local dipole fields on the fields which would exist in the absence of the hole. Thus there is a magnetic dipole in the plane of the hole when there is a tangential magnetic field and an electric dipole normal to the plane of the hole when there is a normal electric field. Note that the dipole fields on opposite sides of the sheet are in antiphase with each other. The dipole moments are proportional to the strengths of the exciting field so that

$$j = \alpha_m H_0$$
$$p = \epsilon_0 \alpha_e E_0 \qquad (9.49)$$

The theory of coupling by small holes is beyond the scope of this book. It is to be found in H. Bethe, Theory of Diffraction by Small Holes, *Phys. Rev.* Vol. 66, pp. 163–182, (1944). Information about the polarizabilities of holes of other shapes is given by Matthaei *et al.* in *Microwave Filters, Impedance-Matching Networks, and Coupling Structures*, Artech House (1980).

where α_m and α_e are the polarizabilities of the hole and depend upon its size and shape. For a circular hole of radius *r* the polarizabilities are given by

$$\alpha_m = \tfrac{4}{3} r^3 \quad \text{and} \quad \alpha_e = \tfrac{2}{3} r^3 \qquad (9.50)$$

We consider first the case of normal incidence and assume that the surface is perfectly conducting. The electric field must be zero within the metal so $E_r = -E_i$ from Equation (9.37). To maintain the correct direction for S_r in Fig. 9.6 it is necessary to reverse the direction of H_r also. The magnetic fields of the incident and reflected waves then add together so that $H = 2H_i$ at the surface of the metal. It can be shown that the boundary condition for the magnetic field on the surface of the metal is satisfied by a surface current flowing in the metal. The fields inside the enclosure are those of a magnetic dipole whose moment is

$8r^3H_i/3$. All points within the box are in the near field of the dipole and the fields are therefore given by Equations (9.11)–(9.13). The field strength at a point on the axis of the hole and distance d from it is

$$|H| = \frac{2}{3\pi}\frac{r^3}{d^3}|H_i| \tag{9.51}$$

from Equation (9.12). The magnetic screening effectiveness is then

$$\begin{aligned} S_M &= -20\log|H/H_i| \\ &= -20\log(2r^3/3\pi d^3) \end{aligned} \tag{9.52}$$

The electric screening effectiveness, calculated in a similar manner, is

$$\begin{aligned} S_E &= -20\log|E/E_i| \\ &= -20\log\left(\frac{4}{3}\frac{r^3}{d^3}\frac{d}{\lambda}\right) \end{aligned} \tag{9.53}$$

Now d is normally much less than λ so $S_M \ll S_E$ and the coupling is caused entirely by the leakage of the magnetic field into the enclosure.

If the incident wave is travelling parallel to the plane of the hole with its electric field normal to the hole, then the electric dipole moment is $2r^3\epsilon E_i/3$. The near field of this dipole is given by Equations (9.1)–(9.3), so the electric field at the centre of the box is, from Equation (9.2)

$$|E| = \frac{1}{6\pi}\frac{r^3}{d^3}|E_i| \tag{9.54}$$

and the electric screening effectiveness is

$$\begin{aligned} S_E &= -20\log|E/E_i| \\ &= -20\log(r^3/6\pi d^3) \end{aligned} \tag{9.55}$$

Similarly, the magnetic screening effectiveness is

$$\begin{aligned} S_M &= -20\log|H/H_i| \\ &= -20\log\left(\frac{1}{3}\frac{r^3}{d^3}\frac{d}{\lambda}\right) \end{aligned} \tag{9.56}$$

The coupling is entirely electric in this case because $S_E \ll S_M$. For some more general angle of incidence the screening effectiveness could be computed by a combination of these two cases.

Worked Example 9.6

Estimate the screening effectiveness at 1 MHz of an enclosure in the form of a cube of sides 200 mm made from aluminium 0.5 mm thick if there is a hole 10 mm in diameter in the centre of one face.

Solution We will assume that the screening effectiveness of the enclosure is determined entirely by the presence of the hole and calculated at the centre of the box. If the incident wave is normal to the face containing the hole, then from Equations (9.52) and (9.53), $S_M = 92$ dB and $S_E = 145$ dB. This confirms that the coupling is entirely magnetic in this case. If the incident wave is tangential to the face containing the hole, then we obtain $S_E = 104$ dB and $S_M = 157$ dB from

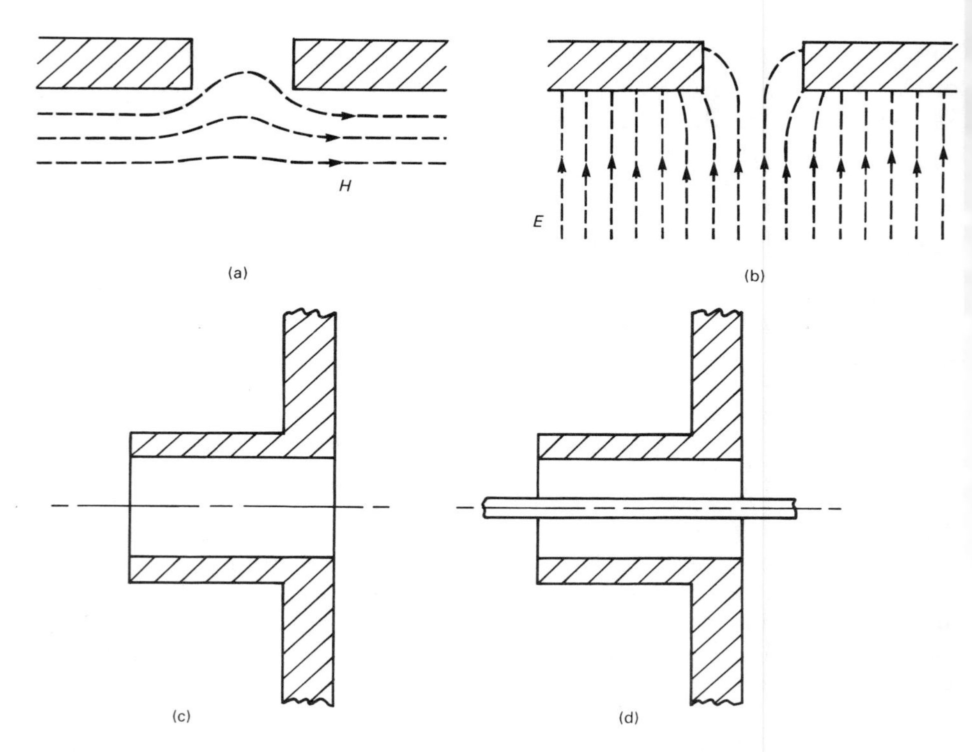

Fig. 9.9 The penetration of (a) a magnetic field and (b) an electric field through a hole in a conducting sheet is affected by the thickness of the sheet. The effective thickness of the sheet can be increased by adding a tube to it as in (c) and (d).

Equations (9.55) and (9.56). These figures can be no more than estimates of the screening effectiveness because the fields of the dipoles are altered by the walls of the box and its contents. Different results would have been obtained if another point within the box had been chosen as the reference point.

The analysis given above has assumed that the screen is very thin. If the thickness of the screen is comparable with the width of the hole, then, as shown in Figs. 9.9(a) and 9.9(b), the fringing fields cannot penetrate through the hole as easily and the screening effectiveness is increased. The screening effectiveness can be increased still further by adding a tube to the hole, as shown in Fig. 9.9(c), if this is possible given the purpose for which the hole exists. The reason for this is that electromagnetic waves cannot propagate down a tube whose diameter is much less than the free-space wavelength. They decay exponentially along the length of the tube.

For a fuller discussion of this subject see the section on waveguides in R.G. Carter.*

The situation is, however, quite different if a wire passes through the tube, as shown in Fig. 9.9(d), because the combination then forms a coaxial transmission line which can propagate waves at all frequencies without attenuation.

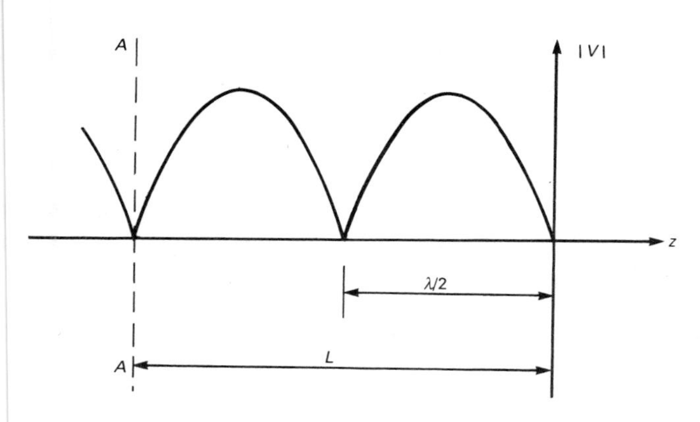

Fig. 9.10 A standing wave on a transmission line.

Electromagnetic resonances

We saw, in Chapter 7, that when waves on a transmission line are reflected by a short circuit the result is a standing wave, as shown in Fig. 9.10. If a second short circuit is placed at one of the nulls of the standing wave such as A–A the wave is unchanged. It follows that a standing wave can exist in the region between two short circuits subject to the condition

$$L = N\lambda/2 \tag{9.57}$$

where N is an integer. Equation (9.57) shows that a standing wave can only exist at one of a set of discrete frequencies. If the section of line is excited at one of those frequencies, then the amplitude of the standing wave is such that the power input is exactly balanced by the losses in the line. The section of line is then **resonant.** If the losses are small, then the wave amplitude can be very high with only a small input power. Evidently, if a screened enclosure can become resonant, then a very small leakage of electromagnetic power into it can produce large internal fields. Alternatively, the resonance maybe excited by a source within the enclosure.

To show how a conducting box can exhibit electromagnetic resonances, we will consider the rectangular box shown in Fig. 9.11. To simplify the analysis we

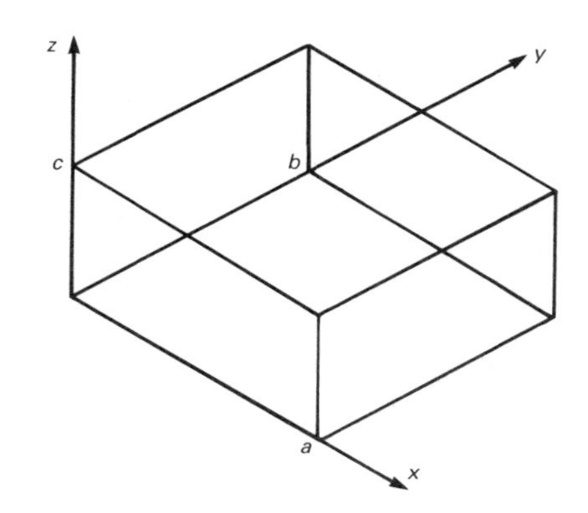

Fig. 9.11 A rectangular conducting box.

will suppose that $\boldsymbol{E}$ has only a z-component and that it varies sinusoidally in the x- and y-directions. Now E_z must be zero on all the vertical walls of the box, so let us suppose that

$$E_z = E_0 \sin(m\pi x/a)\sin(n\pi y/b)\exp(\mathrm{j}\omega t) \tag{9.58}$$

where E_0 is a constant and $m, n = 1, 2, 3, \ldots$. To prove that Equation (9.58) does indeed represent a resonance of the box it is sufficient to show that $\boldsymbol{E}$ satisfies Maxwell's equations.

The magnetic field within the box can be found by substituting for E_z in Equation (8.27), making use of Equation (8.16). The results are

$$(\nabla \wedge \boldsymbol{E})_x = \frac{\partial E_z}{\partial y} \tag{9.59}$$

$$H_x = \mathrm{j}\frac{E_0}{\mu_0\omega}\frac{n\pi}{b}\sin\left(\frac{m\pi x}{a}\right)\cos\left(\frac{n\pi y}{b}\right)\exp(\mathrm{j}\omega t) \tag{9.60}$$

$$(\nabla \wedge \boldsymbol{E})_y = -\frac{\partial E_z}{\partial x} \tag{9.61}$$

$$H_y = -\mathrm{j}\frac{E_0}{\mu_0\omega}\frac{m\pi}{a}\cos\left(\frac{m\pi x}{a}\right)\sin\left(\frac{n\pi y}{b}\right)\exp(\mathrm{j}\omega t) \tag{9.62}$$

Substituting for H_x and H_y in Equation (8.26), we find that $\nabla \wedge \boldsymbol{H}$ only has a z-component given by

$$\begin{aligned}(\nabla \wedge \boldsymbol{H})_z &= \frac{\partial H_y}{\partial x} - \frac{\partial H_x}{\partial y}\\ &= \mathrm{j}\frac{E_0}{\mu_0\omega}\left[\left(\frac{m\pi}{a}\right)^2 + \left(\frac{n\pi}{b}\right)^2\right]\sin\left(\frac{m\pi x}{a}\right)\sin\left(\frac{n\pi y}{b}\right)\exp(\mathrm{j}\omega t)\end{aligned} \tag{9.63}$$

This expression has the same variation with x, y, and t as does E_z and therefore the fields satisfy Maxwell's equations provided that

$$\mathrm{j}\omega\epsilon E_0 = \mathrm{j}\frac{E_0}{\mu_0\omega}\left[\left(\frac{m\pi}{a}\right)^2 + \left(\frac{n\pi}{b}\right)^2\right] \tag{9.64}$$

That is

$$\left(\frac{\omega}{v_\mathrm{p}}\right)^2 = \left(\frac{m\pi}{a}\right)^2 + \left(\frac{n\pi}{b}\right)^2 \tag{9.65}$$

Thus a set of resonances exists whose frequencies are given by Equation (9.65). A more complete analysis of this problem shows that the lowest possible resonant frequency is that for which $m = n = 1$ when a and b are the two longest dimensions of the box. The field pattern for this resonance, deduced from Equations (9.58), (9.60) and (9.62), is shown in Fig. 9.12. Note that the electric and magnetic fields are in phase quadrature, so electromagnetic energy is stored in them but there is no net power flow.

Power can be coupled into the box via holes in the top or bottom (electric dipoles) or in the sides (magnetic dipoles). It is also possible for the resonance to be excited by suitably placed electric or magnetic dipoles anywhere within it. Th

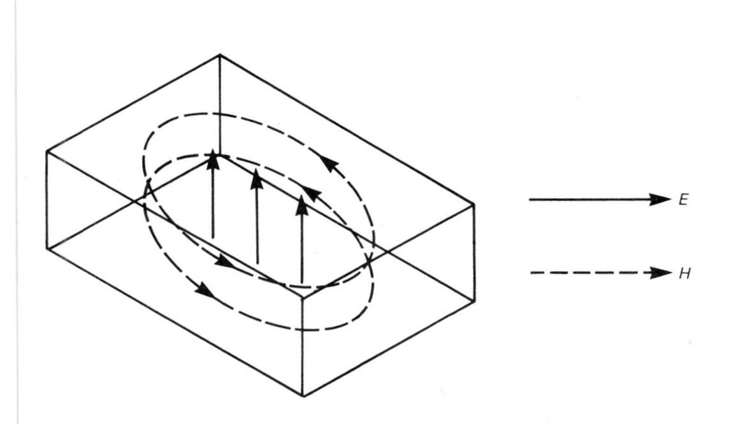

Fig. 9.12 The electric and magnetic fields of the lowest resonant mode of a rectangular conducting box.

requirement for coupling to an electric dipole is that the dipole must be at a point in the box at which there is appreciable electric field and must be arranged so that its field has a component in the same direction as the field of the resonance. A similar argument applies to the magnetic field. Note that, if the resonance is excited by a circuit inside the box, there may be strong coupling between different parts of the circuit. In addition, appreciable electromagnetic energy may escape through any holes and interfere with other circuits nearby.

Further information about electromagnetic resonances can be found in R.G. Carter.*

Worked Example 9.7

Calculate the lowest resonant frequency of a cube of side 200 mm when it is empty and when it is filled with a dielectric material whose relative permittivity is 4.

Solution The frequencies are calculated from Equation (9.65). When the box is empty $v_p = 0.3 \times 10^9$ m s^{-1}. For the lowest resonance $m = n = 1$, so

$$f = \frac{0.3 \times 10^9}{2}\left[\frac{1}{(0.2)^2} + \frac{1}{(0.2)^2}\right]^{1/2} = 1.06 \text{ GHz}$$

When the box is filled with dielectric the phase velocity is halved and the resonant frequency is 530 MHz.

Calculations like the one above can reveal whether the excitation of resonances may be a problem. It is not possible to calculate the exact frequencies of the resonances because they are affected by the contents of the box. Lossy material, placed inside the box, can be used to damp out resonances.

The effect of resonance on screening effectiveness

We noted earlier that the amplitudes of the fields in a resonant enclosure are determined by the balance between the rate at which power is being fed into them and the rate at which it is being dissipated by resistive losses. The relationship between the energy stored in the fields and the rate of energy loss is given by the ***Q*-factor**

$$Q = \frac{2\pi \times \text{Stored Energy}}{\text{Energy dissipated per cycle}} \tag{9.66}$$

The derivation of Equation (9.66) can be found in R.G. Carter.*

An empty closed metal box typically has a Q-factor of a few hundred at a frequency of 1 GHz. This figure is reduced by the presence of any lossy material within the box or by the radiation of energy outwards through holes in its walls.

For the rectangular box considered above the stored energy can be obtained by inegrating the energy density in the peak electric field over the volume of the box. Thus for the lowest resonance the maximum stored electric energy is

$$\begin{aligned} W_E &= \frac{\epsilon_0}{2}\int_0^a\int_0^b\int_0^c |E|^2 \, dx\, dy\, dz \\ &= \frac{\epsilon_0}{2}E_0^2\int_0^a\int_0^b\int_0^c \sin^2(\pi x/a)\sin^2(\pi y/b)\, dx\, dy\, dz \\ &= \frac{abc}{8}\epsilon_0 E_0^2 \end{aligned} \tag{9.67}$$

An electromagnetic resonator, like any other oscillating system such as a pendulum, stores both kinetic and potential energy. Energy is transferred back and forth between the two during the oscillation and the total energy is constant. The total stored energy may, therefore, be obtained by calculating the maximum value of either since at the same instant the other is zero. Thus Equation (9.67) gives both the maximum stored electric potential energy and the total stored energy within the box.

Now, if the power flow into the box is P, then the energy dissipated per cycle in the steady state is

$$W_L = 2\pi P/\omega \tag{9.68}$$

So, from Equations (9.66), (9.67) and (9.68) we have

$$\begin{aligned} P &= \omega W_L/2\pi = \omega W_E/Q \\ &= (\epsilon_0 \omega abc/8Q)E_0^2 \end{aligned} \tag{9.69}$$

This power must equal to the power radiated into the box by the holes in it. Now it is only the far fields of electric and magnetic dipoles which represent power flow. The near fields store reactive energy so they may tune the resonance slightly but they do not couple power into it. It should also be noted that, though the far-field terms are much smaller than those of the near field close to the dipole, they are not zero and the net radial power flow is the same over any spherical surface centred on the dipole. The power radiated into the box by an electric dipole is therefore the integral over a hemisphere, that is

$$P_E = \frac{\omega^2 k_0^2 p^2 Z_w}{24\pi} \tag{9.70}$$

from Equation (9.7). Substituting for the dipole moment from Equation (9.49), we obtain

$$P_E = \frac{\epsilon_0 v_p}{54\pi}k_0^4 r^6 E_i^2 \tag{9.71}$$

Equating this with the power dissipation in the box from Equation (9.69) gives

$$\left(\frac{E_0}{E_i}\right)^2 = \frac{8Q}{54\pi}\frac{k_0^3 r^6}{abc} \tag{9.72}$$

so that the electric screening effectiveness is

$$S_E = -10\log\left(\frac{8Q}{54\pi}\frac{k_0^3 r^6}{abc}\right) \tag{9.73}$$

In the same way, the power radiated by a magnetic dipole is

$$\begin{aligned} P_M &= \frac{1}{24\pi}(Z_w k_0^4 j_m^2) \\ &= \frac{8}{27\pi}\sqrt{\frac{\mu_0}{\epsilon_0}}\, k_0^4 r^6 H_i^2 \end{aligned} \tag{9.74}$$

But $H_i = E_i/Z_w$, so

$$P_M = \frac{8}{27\pi}\sqrt{\frac{\epsilon_0}{\mu_0}}\, k_0^4 r^6 E_i^2 \tag{9.75}$$

Combining Equations (9.69) and (9.75), we obtain

$$\left(\frac{E_0}{E_i}\right)^2 = \frac{64Q}{27\pi}\frac{k_0^3 r^6}{abc} \tag{9.76}$$

so that the magnetic screening effectiveness is

$$S_M = 10\log\left(\frac{64Q}{27\pi}\frac{k_0^3 r^5}{abc}\right) \tag{9.77}$$

This equation has been derived using the value of the electric field at the centre of the box. An alternative equation could have been derived in terms of the magnetic field at the walls.

Worked Example 9.8

Estimate the screening effectiveness of a box 450 × 450 × 150 mm at its lowest resonant frequency if it has a Q-factor of 100 and a hole 10 mm in diameter is cut in the centre of one of its sides.

Solution The lowest resonant frequency of the box is 471 MHz, from Equation (9.65), so $k_0 = 2\pi \times 471/300 = 9.86\ \text{m}^{-1}$. We assume that the leakage of energy directly through the walls is negligible. If a plane electromagnetic wave is incident normally on the face of the box containing the hole then the hole is excited as a magnetic dipole. Then, from Equation (9.77), the magnetic screening effectiveness is 79 dB. If the wave is tangential to the face of the box containing the hole then the electric screening effectiveness is 86 dB from Equation (9.73). The effect of the resonance is shown by comparing these with the corresponding figures derived from Equations (9.52) and (9.55), namely 137 dB and 143 dB. Thus, in the case considered, the resonant fields are much greater than the near fields of the apertures. It should be noted carefully that these figures have been derived by considering the field strength at the centre of the cavity. A different result would have been obtained if another point had been chosen.

Summary

In this chapter we have considered some of the factors affecting the coupling between electronic circuits at radio frequencies and their screening by metal enclosures. The aim has been to provide methods of making estimates of electromagnetic screening effectiveness for emc purposes. The near and far fields of electric and magnetic dipoles were introduced to show how it is possible for circuit elements to radiate or pick up electromagnetic waves. Next we considered the propagation of electromagnetic waves in conducting materials and showed that their effects are confined to a thin surface layer of the material. It was noted in passing that the resistance of circuit components varies with frequency as a result of the skin effect. The attenuation of waves by a conducting sheet is very great provided that the thickness of the material is several times the skin depth. Thus good screening against radiofrequency fields can be provided by closed conducting enclosures. The effectiveness of these enclosures is reduced by the presence of holes in them or by electromagnetic resonances. Small holes can be represented by equivalent dipoles excited by the external electromagnetic fields. Under non-resonant conditions the fields within an enclosure can be estimated from the near-field equations for these dipoles. When the enclosure is resonant the fields inside it can be much greater. Their magnitudes are determined by the balance between the power flow into the enclosure and the rate of dissipation of energy in resistive losses.

Problems

9.1 Estimate the maximum voltage induced in a piece of wire 20 mm long and in a loop of wire whose area is 100 mm^2 when they are placed 100 mm from an electric dipole 10 mm long and 1 mm in diameter which is excited by a low-impedance 10 V source at a frequency of 100 kHz.

9.2 Repeat Problem 9.1 with the source replaced by a circular loop 20 mm in diameter made of wire 1 mm thick and driven by a 1 A current source at 100 kHz.

9.3 Calculate the skin depth in copper, iron and nickel at frequencies of 50 Hz, 1 kHz and 1 MHz.

9.4 Calculate the resistance per metre of copper wires 2 mm and 0.2 mm in diameter at frequencies of 50 Hz, 1 kHz and 1 MHz.

9.5 Calculate the screening effectiveness of a sheet of aluminium foil 0.05 mm thick at frequencies of 1 MHz, 10 MHz and 100 MHz.

9.6 A screening enclosure 400 × 400 × 200 mm has a circular hole 20 mm in diameter cut in the centre of one of its broad faces. Estimate the minimum screening effectiveness of the enclosure at 10 kHz, 100 kHz and 1 MHz, assuming that the screening effectiveness was very high before the hole was made.

9.7 Calculate the lowest resonant frequency of rectangular boxes having the following dimensions

(a) 50 × 70 × 30 mm
(b) 200 × 300 × 100 mm
(c) 500 × 400 × 350 mm
when they are empty and when they are filled with a material whose effective permittivity is 15.

9.8 Estimate the screening effectiveness of the enclosure described in Problem 9.6 at its lowest resonant frequency if its Q-factor is 500 and if its Q-factor is reduced to 50 by the inclusion of lossy material.

Bibliography

Electromagnetism

Elementary

Bolton, B., *Electromagnetism and its Applications: an introduction*, Van Nostrand Reinhold (1980).
Compton, A.J., *Basic Electromagnetism and its Applications*, Van Nostrand Reinhold (1986).

Intermediate

Bleaney, B.I. and Bleaney, B., *Electricity and Magnetism* (3rd edn), Oxford University Press (1976).
Carter, R.G., *Electromagnetic Waves: Microwave Components and Devices*, Chapman and Hall (1990).
Carter, G.W., *The Electromagnetic Field in its Engineering Aspects* (2nd edn), Longman (1967).

Advanced

Reitz, J.R. and Milford, F.J., *Foundations of Electromagnetic Theory* (3rd edn), Addison-Wesley (1979).

Solution of field problems

Binns, K.J. and Lawrenson, P.J., *Analysis and Computation of Electric and Magnetic Field Problems*, Pergamon (1963).
Silvester, P.P. and Ferrari, R.L., *Finite Elements for Electrical Engineers*, Cambridge University Press (1983).
Smythe, W.R., *Static and Dynamic Electricity*, McGraw-Hill (1939).

Energy methods

Hammond, P., *Energy Methods in Electromagnetism*, Oxford University Press (1981).

Applications of electromagnetism

Electrostatics

Bright, A.W., Corbett, R.P. and Hughes, J.F., *Electrostatics*, Oxford University Press (1978).
Moore, A.D. (ed.), *Electrostatics and its Applications*, Interscience (1973).

Magnetism

Parker, R.J. and Studders, R.J., *Permanent Magnets and their Applications*, Wiley (1962).

Wright, W. and McCaig, M., *Permanent Magnets*, Oxford University Press (1977).

Electric and magnetic devices

Bar-Lev, A., *Semiconductors and Electronic Devices* (2nd edn), Prentice Hall (1984).

Dummer, G.W.A. and Nordenberg, H.N., *Fixed and Variable Capacitors*, McGraw-Hill (1960).

Grossner, N.R., *Transformers for Electronic Circuits*, McGraw-Hill (1967).

Sangwine, S.J., *Electronic Components and Technology: Engineering Applications*, Van Nostrand Reinhold (UK) (1987).

Slemon, G.R., *Magneto-electric Devices*, Wiley (1966).

Spangenburg, K., *Vacuum Tubes*, McGraw-Hill (1948).

Electromagnetic interference

Freeman, E.R. and Sachs, H.M., *Electromagnetic Compatibility Design Guide*, Artech House (1982).

Keiser, B.E., *Principles of Electromagnetic Compatibility*, Artech House (1983).

Morrison, R., *Grounding and Shielding Techniques in Instrumentation*, Wiley (1977).

Walker, C.S., *Capacitance, Inductance and Crosstalk Analysis*, Artech House (1990).

Properties of materials

Dummer, G.W.A., *Materials for Conductive and Resistive Functions*, Hayden Book Co. (1970).

Heck, C., *Magnetic Materials and their Applications*, Butterworths (1974).

Sillars, R.W., *Electrical Insulating Materials and their Application*, Peter Peregrinus (1973).

General reference

Fink, D.G. and Christiansen, D., *Electronic Engineers' Handbook* (2nd edn), McGraw-Hill (1982).

Mazda, F. (ed.), *Electronic Engineer's Reference Book* (6th edn), Butterworths (1990).

Note

The books in this list should provide you with the means of obtaining further information on any part of the theory of electromagnetism discussed in this book. They should also provide a way into the professional literature dealing with the applications of electromagnetism. Some of the older books are now out of print; they have been included because of their lasting value or because no more recent books exist on those subjects.

Appendix

Physical constants

Primary electric constant	8.854×10^{-12}	$F\ m^{-1}$
Primary magnetic constant	$4\pi \times 10^{-7}$	$H\ m^{-1}$
Velocity of light in vacuum	2.998×10^{8}	$m\ s^{-1}$
Wave impedance of free space	376.7	Ω
Charge on the electron	-1.602×10^{-19}	C
Rest mass of the electron	9.108×10^{-31}	kg
Charge/mass ratio of the electron	-1.759×10^{11}	$C\ kg^{-1}$

Properties of dielectric materials

	Relative permittivity
Alumina 99.5%	10
Alumina 96%	9
Barium titanate	1200
Beryllia	6.6
Epoxy resin	3.5
Ferrites	13–16
Fused quartz	3.8
GaAs (high resistivity)	13
Nylon	3.1
Paraffin wax	2.25
Perspex	2.6
Polystrene	2.54
Polystyrene foam	1.05
Polythene	2.25
PTFE (Teflon)	2.08

Properties of conductors

	Conductivity ($S\ m^{-1}$)
Aluminium	3.5×10^{7}
Brass	1.1×10^{7}
Copper	5.7×10^{7}
Distilled water	2×10^{-4}
Ferrite (typical)	10^{-2}
Fresh water	10^{-3}
Gold	4.1×10^{7}
Iron	0.97×10^{7}
Nickel	1.28×10^{7}
Sea water	4

Silver	6.1×10^7
Steel	0.57×10^7

Properties of ferromagnetic materials

	μ_r	Saturation magnetism (B_{sat}) (T)
Feroxcube 3	1 500	0.2
Mild steel	2 000	1.4
Mumetal	80 000	0.8
Nickel	600	
Silicon iron	7 000	1.3

Summary of vector formulae in Cartesian coordinates

$$\begin{aligned}
\boldsymbol{a} &= a_x\hat{\boldsymbol{x}} + a_y\hat{\boldsymbol{y}} + a_z\hat{\boldsymbol{z}} \\
\boldsymbol{a}\cdot\boldsymbol{b} &= a_xb_x + a_yb_y + a_zb_z \\
\boldsymbol{a}\wedge\boldsymbol{b} &= (a_yb_z - a_zb_y)\hat{\boldsymbol{x}} + (a_zb_x - a_xb_z)\hat{\boldsymbol{y}} + (a_xb_y - a_yb_x)\hat{\boldsymbol{z}} \\
\nabla &= \frac{\partial}{\partial x}\hat{\boldsymbol{x}} + \frac{\partial}{\partial y}\hat{\boldsymbol{y}} + \frac{\partial}{\partial z}\hat{\boldsymbol{z}} \\
\text{grad } V &= \nabla V \\
&= \frac{\partial V}{\partial x}\hat{\boldsymbol{x}} + \frac{\partial V}{\partial y}\hat{\boldsymbol{y}} + \frac{\partial V}{\partial z}\hat{\boldsymbol{z}} \\
\text{div } \boldsymbol{a} &= \nabla\cdot\boldsymbol{a} \\
&= \frac{\partial a_x}{\partial x} + \frac{\partial a_y}{\partial y} + \frac{\partial a_z}{\partial z} \\
\text{curl } \boldsymbol{a} &= \nabla\wedge\boldsymbol{a} \\
&= \left\{\frac{\partial a_z}{\partial y} - \frac{\partial a_y}{\partial z}\right\}\hat{\boldsymbol{x}} + \left\{\frac{\partial a_x}{\partial z} - \frac{\partial a_z}{\partial x}\right\}\hat{\boldsymbol{y}} + \left\{\frac{\partial a_y}{\partial x} - \frac{\partial a_x}{\partial y}\right\}\hat{\boldsymbol{z}} \\
\nabla^2 V &= \frac{\partial^2 V}{\partial x^2} + \frac{\partial^2 V}{\partial y^2} + \frac{\partial^2 V}{\partial z^2} \\
\nabla\wedge(\nabla\wedge\boldsymbol{a}) &= \nabla(\nabla\cdot\boldsymbol{a}) - \nabla^2\boldsymbol{a} \\
\nabla^2\boldsymbol{a} &= \hat{\boldsymbol{x}}\nabla^2 a_x + \hat{\boldsymbol{y}}\nabla^2 a_y + \hat{\boldsymbol{z}}\nabla^2 a_z
\end{aligned}$$

Summary of the principal formulae of electromagnetism

Inverse square law of electrostatic force

$$\boldsymbol{F} = \frac{Q_1 Q_2}{4\pi\epsilon_0 r^2}\hat{\boldsymbol{r}}$$

Relationship between $\boldsymbol{E}$ and $\boldsymbol{D}$

$$\boldsymbol{D} = \epsilon\boldsymbol{E} = \epsilon_0\epsilon_r\boldsymbol{E}$$

Gauss' theorem

$$\oint\!\!\!\oint \boldsymbol{D}\cdot\mathbf{d}\boldsymbol{A} = \iiint \rho\,\mathrm{d}v$$

$$\nabla\cdot\boldsymbol{D} = \rho$$

Electrostatic potential difference

$$V_B - V_A = -\int_A^B \boldsymbol{E} \cdot \mathbf{d}\boldsymbol{l}$$

$$\boldsymbol{E} = -\nabla V$$

Poisson's equation

$$\nabla^2 V = -\rho/\epsilon$$

Energy density in an electric field

$$w = \tfrac{1}{2}\boldsymbol{E}\cdot\boldsymbol{D}$$

Ohm's law

$$\boldsymbol{J} = \sigma \boldsymbol{E}$$
$$\boldsymbol{E} = \rho \boldsymbol{J}$$

Power dissipated per unit volume

$$p = \boldsymbol{E}\cdot\boldsymbol{J}$$

Continuity equation

$$\oiint \boldsymbol{J}\cdot\mathbf{d}\boldsymbol{A} = -\frac{\partial}{\partial t}\iiint \rho \, \mathrm{d}v$$

$$\nabla\cdot\boldsymbol{J} = -\frac{\partial \rho}{\partial t}$$

Electromotive force

$$\mathcal{E} = -\oint \boldsymbol{E}\cdot\mathbf{d}\boldsymbol{l}$$

Law of force between moving charges

$$\boldsymbol{F} = \frac{Q_1 Q_2}{4\pi\epsilon_0 r^2}\hat{\boldsymbol{r}} + \frac{\mu_0 Q_1 Q_2}{4\pi r^2}\left(\boldsymbol{v}_2 \wedge (\boldsymbol{v}_1 \wedge \hat{\boldsymbol{r}})\right)$$

Force on a moving charge

$$\boldsymbol{F} = Q\left(\boldsymbol{E} + (\boldsymbol{v}\wedge\boldsymbol{B})\right)$$

Force on a current-carrying conductor

$$\boldsymbol{F} = I(\mathbf{d}\boldsymbol{l}\wedge\boldsymbol{B})$$

Biot–Savart law

$$\boldsymbol{B} = \frac{\mu_0 I}{4\pi}\oint \frac{\mathbf{d}\boldsymbol{l}\wedge\hat{\boldsymbol{r}}}{r^2}$$

Relationship between $\boldsymbol{B}$ and $\boldsymbol{H}$

$$\boldsymbol{B} = \mu\boldsymbol{H} = \mu_0\mu_r\boldsymbol{H}$$

Magnetic circuit law as modified by Maxwell

$$\oint \boldsymbol{H}\cdot\mathbf{d}\boldsymbol{l} = \iint\left(\boldsymbol{J} + \frac{\partial \boldsymbol{D}}{\partial t}\right)\cdot\mathbf{d}\boldsymbol{A}$$

$$\nabla\wedge\boldsymbol{H} = \boldsymbol{J} + \frac{\partial \boldsymbol{D}}{\partial t}$$

Conservation of magnetic flux

$$\oiint \boldsymbol{B} \cdot \mathbf{d}\boldsymbol{A} = 0$$
$$\nabla \cdot \boldsymbol{B} = 0$$

Faraday's law of induction

$$\oint \boldsymbol{E} \cdot \mathbf{d}\boldsymbol{l} = -\frac{\partial}{\partial t} \iint \boldsymbol{B} \cdot \mathbf{d}\boldsymbol{A}$$

$$\nabla \wedge \boldsymbol{E} = -\frac{\partial \boldsymbol{B}}{\partial t}$$

Energy density in a magnetic field

$$w = \tfrac{1}{2} \boldsymbol{B} \cdot \boldsymbol{H}$$

The Poynting vector

$$\boldsymbol{P} = \boldsymbol{E} \wedge \boldsymbol{H}$$

Answers to problems

1.1 *Hint*: Construct a Gaussian surface in the shape of a thin flat box enclosing part of the surface of the metal with the broad faces of the box parallel to the surface.

1.2 $V_{axis} = (q/4\pi\epsilon_0)\,[1 + 2\ln(a/b)]$
$u_0 = 58.2 \times 10^6\ \mathrm{m\,s^{-1}}$
$V_{axis} = -368.4\ \mathrm{V}$

1.3 $F_r = 4.95 \times 10^{-15}$ N when
$b = 10$ mm
$z_{max} \sim 110$ mm

1.5 $V_{max} = 764$ kV
$\sigma_{max} = 26.6 \times 10^{-6}\ \mathrm{C\,m^{-2}}$
Hint: Use the method of images.

1.6 Symmetrically with a positive charge in the third quadrant and negative charges in the second and fourth quadrants. For 60°, five image charges, alternately positive and negative, are needed.
The method will not work for angles which are not exact divisions of 360°.
Solve for angles 90°, 60°, 45°, 30° and 15° and interpolate.

1.7 The extreme field strengths are at the ends of the diagonals.
At the inner end $E = 4100\ \mathrm{V\,m^{-1}}$.
At the outer end $E = 280\ \mathrm{V\,m^{-1}}$.
Hint: Solve the problem using the finite difference method.

1.8 *Hint*: In the steady state the current density is constant, but the charge density, potential and electron velocity all depend on the position.

2.1 $V_{max} = 150$ V
If the sheet completely fills the space $V_{max} = 3000$ V.

2.2 *Hint*: Use the result of Worked Example 1.3 to find the potential difference between the rods when there are charges $\pm q$ per unit length on them.

2.3 $C = 10\ \mathrm{pF\,m^{-1}}$
Hint: Use the method of images to calculate the capacitance between each wire and the plane.

2.4 $b = a\,(\epsilon_1 E_1)/(\epsilon_2 E_2)$

2.5 The radius of the edge of the plates must vary as $(1 + \theta/50)^{1.5}$ where θ is in degrees.

2.6 By finite difference methods $C = 15.7\epsilon_0\ \mathrm{F\,m^{-1}}$.
By energy methods $15.6\epsilon_0 \le C \le 16.0\epsilon_0$.

2.7 *Hint*: Find the potential drop across the two parts of the depletion layer and add.

2.8 $C_j = (\epsilon q/2V)^{1/2} N_D^{1/2} (1 + N_D/N_A)^{-1/2}$

2.9 $V_{max} = 60$ V

2.10 $\sigma_{max} = 0.32\ \mathrm{C\,m^{-2}}$
Proportion ionized 1.07%

3.1 $v_D = 79.0 \times 10^{-6}\ \mathrm{m\,s^{-1}}$
Distance travelled $= 0.71 \times 10^{-6}$ m
Power dissipated $= 71\ \mathrm{mW\,m^{-1}}$

3.2 $R = (\rho/2\pi)\ln(b/a)$

3.3 If $V_{DS} \ll V_{GS}$
$I_{DS} = (2w/\rho L)\,(a - \sqrt{V_{GS}/V_p})\,V_{DS}$
Otherwise

$$I_{DS} = (2wa/\rho L)\left\{ V_{DS} - \frac{2}{3}\frac{(V_{DS} - V_{GS})^{3/2}}{\sqrt{V_p}} + \frac{2}{3}\frac{(V_{GS})^{3/2}}{\sqrt{V_p}} \right\}$$

where $t = a$ when $\sqrt{V + V_{GS}} = \sqrt{V_p}$

3.4 The corner is equivalent to a straight section of length $0.98a$.

3.5 The change in the resistance is 6% and is independent of the position of the hole.

4.1 $$B = (\mu_0 I/2R)\,\frac{\sin(\pi/2n)}{(\pi/2n)}$$

4.2 $$B = \frac{2\sqrt{2}\mu_0 I}{\pi}\,\frac{a^2}{\sqrt{(a^2 + 2z^2)(a^2 + 4z^2)}}$$

4.3 $B_z = 5\mu_0 I/(4\pi a)$

4.4 (a) $B = (\mu_0 nI/2a)(\cos\beta - \cos\alpha)$
(b) $B = (\mu_0 nI/\sqrt{2}\,a)$
(c) $B = \mu_0 nI$

4.5 $z = a$
Hint: Find a general expression for B on the axis of the coils and adjust d to make dB/dz, d^2B/dz^2, etc., zero to the highest possible order. This is equivalent to making the Taylor expansion of the field take the form $B = a_0 + a_n z^n + \ldots$ where n is as large as possible.

4.6 $B = \mu_0 nI$ (n turns per unit length)

4.7 *Hint*: Consider the conservation of

energy and angular momentum. The solution to this problem is given by Bleaney and Bleaney.*

4.8 *Hint*: Write down the equation of angular motion and an expression for Φ in terms of B_z. Then, using div $B = 0$ show that

$$\frac{\mathrm{d}}{\mathrm{d}t}(r^2\theta) = \frac{\eta}{2\pi}\frac{\mathrm{d}\Phi}{\mathrm{d}t}$$

4.9 $$B^2 = \frac{2I}{\eta\pi b^2\epsilon_0\sqrt{(2\eta V)}}$$

5.1 Assuming that there are N turns in each coil, and that the current in each coil is I, the field is uniform to about 1 part in 10^5 and $B = \mu_0 NI/a$. If the coil spacing is doubled the maximum and minimum values of the field are $0.5013\ \mu_0 NI/a$ and $0.4985\mu_0 NI/a$.
Hint: Tabulate the field of one coil at intervals of $a/2$ and sum the fields of the coils and their images in the planes. If you have access to a computer you will find that that makes the task much easier.

5.2 $I_{max} = 86$ mA (rms)

5.3 670 turns

5.4 (a) $B = 1.12$ T
(b) $B = 1.23$ T

5.5 $B_{air} = 0.55$ Tesla
If the magnet dimensions are optimized the saving in volume is 3% which probably does not justify a design change.

5.6 The weight of the Alnico magnet is 1.83 kg and that of the $SmCo_5$ magnet is 0.17 kg. In the second case additional iron would be needed to conduct the flux from the smaller magnet offsetting some of the saving in weight.
Hint: Both magnets must supply the same flux at the same magneto-motive force.

6.1 $M = 395$ pH
The flux linkage varies as the cosine of the angle between the planes of the coils.
Hint: Use the reciprocity of mutal inductance. Find the flux generated by the large coil which is linked to the smaller one.

6.2 $V = 0.51$ V (rms)
The answer is independent of the position of the wire within the toroid.

6.3 $V = 0.24\ \mu$V (rms) when the plane of the circuit coincides with the one in which the two wires lie.

6.4 $L = 0.23$ mH, $I_{max} = 1.3\ A$

6.5 $L_1 = 16$ mH, $L_2 = 400$ mH, $M = 80$ mH

6.6 $L = 79.4 \pm 0.8$ pH m^{-1}

7.1 $\Gamma = \pm\ 0.33$
$Z/Z_0 = 0.5$ or 2.0
Return loss $= -\ 9.5$ dB

7.2 At A the amplitudes of the voltage pulses are $V_s/2$, $V_s/6$, $V_s/18$ etc. At B they are $2V_s/3$, $2V_s/9$ etc. (all pulses are positive).

7.3 $L = 1.0$ mH m^{-1}
$Z_0 = 500\ \Omega$
Pulse amplitude = 10 kV, pulse length = 2.5 μs
Load current = 20 A, power in the pulse = 0.5 J

7.4 At 500 MHz $\lambda = 0.364$ m, $Z_{in} = 33.3 - 0.5j\ \Omega$
At 1 GHz $\lambda = 0.182$ m, $Z_{in} = 75.0 + 2.0j\ \Omega$
At 2 GHz $\lambda = 0.091$ m, $Z_{in} = 74.9 + 4.3j\ \Omega$

7.5 Diameter of the inside conductor = 4.35 mm
The quarter-wave transformer would be 75 mm long and have an inner conductor 3.61 mm in diameter.

7.7 $v_p = 0.2998 \times 10^9$ m s^{-1}
$Z_0 = 359\ \Omega$

7.8 $v_p = 0.15 \times 10^9$ m s^{-1}
$Z_0 = 24.8\ \Omega$

9.1 0.59 mV, 1.22 pV

9.2 39.4 μV, 1.97 μV

9.3 Copper: 9.4 mm, 2.1 mm, 0.066 mm
Iron: 0.27 mm, 0.061 mm, 1.93 μm
Nickel: 0.81 mm, 0.18 mm, 5.7 μm

9.4 2 mm dia: 6.3 mΩ/m, 14.9 mΩ/m, 43.8 mΩ/m
0.2mm dia: 0.56 Ω/m, 0.56 Ω/m, 0.63 Ω/m

9.5 111 dB, 122 dB, 157 dB

9.6 73 dB in all cases.

9.7 Empty: 3.68 GHz, 901 MHz, 480 MHz
With dielectric: 950 MHz, 233 MHz, 124 MHz

9.8 Q = 500: $S_E = 60$ dB, $S_M = 48$ dB
Q = 50: $S_E = 70$ dB, $S_M = 58$ dB

Index

Ampere (unit of current), 44
Ampere's Law *see* magnetic circuit law
Ampere-turns, 70

Biot-Savart Law, 57
Boundary conditions
  electric current, 46
  electrostatic, 26
  magnetic, 70
Boundary element method, 39
Breakdown, dielectric, 23
  of air, 13
Brillouin field, 66
Busch's theorem, 66

Calculation
  of capacitance, 31
  of capacitance by energy methods, 37
  of electrostatic potential difference, 7
  of inductance, 98
  of inductance by energy methods, 103
  of resistance, 49
  of resistance by energy methods, 51
Capacitance, 28
  calculation by energy methods, 37
  calculation of, 31
  of coaxial line, 31, 127
  of parallel wires, 40
  of p–n junction, 41
  stray, 28
Cathode ray tube, 19–20
Characteristic impedance of transmission line, 117
Charge
  conservation of, 45
  induced, 12
  mobility of, 44
  of electron, 2, 166
  polarization, 24
Charged particles
  motion in electric and magnetic fields, 63
  motion in electric field, 19
  motion in magnetic fields, 61
Choke *see* inductor
Circuit, magnetic, 74
Coaxial line
  capacitance per unit length, 31, 127
  fields in, 128
  inductance per unit length, 100
  lossy, 47
  power flow in, 129
Coercive force, 79
Coils, Helmholtz, 65
Complex numbers, representation of waves by, 116
Conducting materials, 43
  table of, 166
Conduction
  of electricity, 11, 43
  of magnetic flux, 71
Conduction charges, 12, 43
Conductivity, 44
Conformal mapping, 16
Conservation
  of charge, 45
  of energy, 8
Constitutive relations, 138
Continuity equation, 46
Coulomb (unit of charge), 2
Curl of vector, 136
Current density, 44
Current
  displacement, 134
  eddy, 109
  induced, 90–3
  loop, magnetic field of, 57
  total, 134
  transformer, 111
Cyclotron frequency, 62

Del, vector operator, 7
Demagnetization curve, 83
Demagnetizing field of air gap, 83
Dielectric breakdown, 23
  of air, 13
Dielectric constant *see* relative permittivity
Dielectric materials, 23
  permittivity, 25
  polarization, 24
  table of, 166
Dipole
  electric, 143
  magnetic, 146
Dispersion, of waves, 120
Displacement current, 134
Divergence of vector, 6
Drift velocity, 43, 57
Duality, principle of, 50

Earth loops, 97
Eddy current, 109
  loss, 109–10
Electric constant, primary (ϵ_0), 2, 166
Electric current, power dissipation by, 45
Electric dipole, 143
  equivalent circuit, 146
  input resistance, 145
  moment, 144
Electric field (E), 3
  calculation from potential, 10
  energy storage in, 35
  in presence of currents, 47
  lines, 3
Electric flux, 4
Electric flux density (D), 25
Electric force
  on charge, 3
  inverse square law of, 2
Electric potential
  calculation by line integral, 7
  gradient of, 11
Electric screening effectiveness, 30, 155, 161
Electrolytic tank, 47
Electromagnetic
  compatibility, 142
  induction, 90
    Faraday's Law of, 93
  interference, 29, 96, 142

pulse, nuclear, 29
resonance, 157
Electromagnetic waves
in conducting materials, 148
in free space, 138
power flow in, 139
reflection by conducting surface, 151
Electromotive force, 48, 91–3
Electron
charge of, 2, 166
rest mass of, 166
Electrostatic
force between charges, 2
potential, 7
screening, 29
emf *see* electromotive force
Energy, conservation of, 8
Energy product, 85
Energy storage
in capacitor, 36
in electric field, 35
in inductor, 101
in magnetic field, 101
in magnetized iron, 107
Energy methods
calculation of capacitance by, 37
calculation of inductance by, 103
calculation of resistance by, 51
Equipotential surface
electric, 11
magnetic, 72
Equivalent circuit
of electric dipole, 146
of magnetic dipole, 147
Evershed's criterion, 85

Farad (unit of capacitance), 28
Faraday cage, 30
Faraday's Law of electromagnetic induction, 93
differential form, 136
Ferrites, 110
Ferromagnetism, 68
Finite difference method, 17
Finite element method, 38
Flux conduction, magnetic, 71
Flux
electric, 4
leakage, magnetic, 76
linkage, magnetic, 93
magnetic, 61, 74, 91
tube, 36
Flux density
electric (D), 25
magnetic (B), 56
Force
electric, on charge, 3
electrostatic, between charges, 2
magnetic, between moving charges, 55
magnetic, on moving charge, 61
on current-carrying wire, 64
Fringing of magnetic flux, 75

Gauss' theorem
electrostatic, differential form, 6, 25
electrostatic in dielectric materials, 25
electrostatic in free space, 4
magnetic, 61
Gaussian surface, 4
Gradient
of electric potential, 11
of vector, 11

Hall effect, 63
Hard magnetic materials, 80
Helmholtz coils, 65
Henry (unit of inductance), 95
Hysteresis, 78
loop, 79
loop, minor, 80, 83
loss, 108

Images, method of
electrostatics, 14
magnetism, 72
Impedance
characteristic of transmission line, 117
intrinsic of free space, 139, 166
transformation on transmission line, 124
Induced
charge, 12
current, 90–3
Inductance
calculation of, 98
calculation of by energy methods, 103
mutual, 95
of coaxial line, 100
self, 95
stray, 100
Inductor, 100
energy storage in, 101
toroidal, 102
Initial magnetization curve, 79
Interference, electromagnetic, 29, 96, 142
Intrinsic impedance of free space, 139, 162
Inverse square law of electric force, 2
Iron
energy storage in, 107
magnetic effects of, 67

Joule's Law, 45

Kirchhoff
current law, 46
voltage law, 49

Lamination, transformer core, 110
Laplace's equation, 16, 46, 61
finite difference solution, 17–19
LCRZ analogy, 105
Leakage of magnetic flux, 76
Lenz's Law, 93
Light, velocity of, 127, 139, 166
Line integral, 7, 59, 94
Line of force
electric, 3
magnetic, 72
Loss
eddy current, 109–10
hysteresis, 108
reflection, 152
transmission, 153
Loudspeaker, 84
Lumped components, 114

Magnet, permanent, 83
Magnetic
circuit, 74
constant, primary (μ_0) 55, 166

Magnetic (cont.)
dipole, 146
equivalent circuit, 147
moment, 147
radiation resistance of, 148
effects of Iron, 67
equipotential, 72
pole, 55
Magnetic circuit law
differential form, 135
in free space, 59
in magnetic materials, 70, 74
Maxwell's form, 133
Magnetic field (H), 69
energy storage in, 101
fringing of, 75
line of, 72
of current loop, 57
of square coil, 64
of straight wire, 58
of toroidal coil, 60
Magnetic flux, 61, 74, 91
conduction, 71
density (B), 56
unit of (Tesla), 56
leakage, 76
Magnetic force
between charges, 55
on current-carrying wire, 64
on moving charge, 61
Magnetic materials, 68, 78
hard, 80
permeability of, 69
remanence of, 79
saturation of, 79
soft, 79
susceptibility, 69
table of, 167
Magnetic scalar potential, 60
Magnetic screening effectiveness, 71, 155, 161
Magnetization curve, initial, 79
Magnetomotive force, 75
Magnetron, 65, 89
Mass of electron, 166
Matched transmission line, 118
Matching by quarter-wave transformer, 126
Materials
conducting, 43
table of, 166
dielectric, 23
table of, 166
magnetic, 68, 78
hard, 80
soft, 79
table of, 167
Maxwell's equations, 133, 137
Method of images
electrostatics, 14
magnetism, 72
mmf *see* magneto-motive force
Mobility, of charge, 44
Mutual inductance, 95

Near field
of electric dipole, 143
of magnetic dipole, 147
Numerical methods
boundary element, 39
finite difference, 17
finite element, 38
tubes and slices, 38

Ohm (unit of resistance), 44
Ohm's law, 44
Ohmic heating, 44
Ohms per square, 50

Parasitic
capacitance *see* stray capacitance
inductance *see* stray inductance
Permanent magnets, 83
efficient use of, 85
energy product, 85
stabilization, 84
Permeability, 69
relative, 69
Permittivity, 25
relative, 24
Phase velocity of wave, 116
P–N Junction, capacitance of, 41
Poisson's equation, 16
Polarization of dielectric materials, 24
Pole, magnetic, 55
Potential
electrostatic, 7
unit of (Volt), 8
magnetic scalar, 60
Power flow
in coaxial line, 129
in electromagnetic wave, 139
Poynting vector, 140
Poynting's theorem, 140
Primary electric constant (ϵ_0), 2, 166
Primary magnetic constant (μ_0), 55, 166
Principle
of conservation of energy, 8
of duality, 50
of superposition, 3, 58
Propagation constant of wave, 116
Pulse-forming network, 131
Pulse, nuclear electromagnetic, 29
Pulses
on transmission line, 119
reflection of, 120

Q factor, 159
Quarter-wave transformer, 126

Real electronic components, 110
Reflection
coefficient, voltage, 118
loss, 152
of pulses on transmission line, 120
of waves on transmission line, 117
Relative
permeability, 69
permittivity, 24
Relaxation time, 11
Reluctance, 75
Remanence of magnetic materials, 79
Resistance
calculation of, 49
by energy methods, 51
sheet, 50
Resistivity, 44
Resistor
diffused, integrated circuit, 49
pinch, integrated circuit, 54
Resonance
electromagnetic, 157
Q factor of, 159

Saturation of magnetic materials, 79
Screening
electrostatic, 29

magnetic, 71
radiofrequency, 142
Screening effectiveness
effect of holes on, 153
effect of resonance on, 159
electric, 30, 155, 161
magnetic, 71, 155, 161
Self inductance, 95
Sheet resistance, 50
Siemens (unit of conductance), 44
Skin depth, 149
Skin effect, 150
Soft magnetic materials, 79
Spherical polar coordinates, 143–4
Stabilization of permanent magnets, 84
Stray
capacitance, 28
inductance, 100
Superposition, principle of, 3, 58
Susceptibility, magnetic, 69

TEM wave, 129
Tesla (unit of magnetic flux density), 56
Thevenin voltage, 48
Transformer, 99
core lamination, 110
current, 111
quarter wave, 126
Transmission line, 113
characteristic impedance, 117
dispersion, 120
field of, 127
impedance transformation on, 124
pulses on, 119
quarter wave transformer, 126
reflection of pulses on, 120
reflection of waves on, 117
standing wave on, 119
waves on, 115, 117
Transmission loss, 153
Tube
cathode ray, 19–20
TV picture, 63
Tubes and slices, method of, 38
Tube of flux, 36

Vector
curl of, 136
divergence of, 6
gradient of, 11
Velocity
of light, 127, 139, 166
phase, of wave, 116
drift, 43, 57
Volt (unit of potential difference), 8
Voltage reflection coefficient, 118
Voltage standing wave ratio (VSWR), 119
VSWR *see* voltage standing wave ratio

Wave
phase velocity of, 116
propagation constant, 116
standing, on transmission line, 119
TEM, 129
Wave equation, 115, 139
Wave impedance, 128–9
Wavenumber *see* propagation constant
Waves
dispersion of, 120
electromagnetic,
in conducting materials, 148
in free space, 138
reflection of by conducting surface, 151
on transmission lines, 115, 117
representation by complex numbers, 116
Weber (unit of magnetic flux), 61